Nuclear Physics

Nuclear Physics

Leonel Ware

Larsen & Keller
www.larsen-keller.com

Nuclear Physics
Leonel Ware
ISBN: 978-1-64172-110-3 (Hardback)

Larsen & Keller
Published by Larsen and Keller Education,
5 Penn Plaza,
19th Floor,
New York, NY 10001, USA

Cataloging-in-Publication Data

Nuclear physics / Leonel Ware.
p. cm.
Includes bibliographical references and index.
ISBN 978-1-64172-110-3
1. Nuclear physics. 2. Physics. I. Ware, Leonel.
QC776 .N83 2019
539.7--dc23

For more information regarding Larsen and Keller Education and its products, please visit the publisher's website www.larsen-keller.com

Table of Contents

Preface

Nuclear physics is a field of physics, which studies the nucleus of an atom and its constituents and the interactions between them. The study of the nucleus can be approached from both classical and quantum mechanical theories using the liquid-drop model and nuclear shell model respectively. Interacting boson model and ab initio methods are alternative theories to the study of the nucleus. Some of the areas of study in nuclear physics include nuclear decay, radioactivity, nucleosynthesis, etc. Advances in nuclear physics have brought about advancements in nuclear medicine, nuclear power, nuclear weapons, magnetic resonance imaging and radiocarbon dating. This book is a valuable compilation of topics, ranging from the basic to the most complex theories and principles in the field of nuclear physics. Different approaches, evaluations and methodologies have been included herein. This book, with its detailed analyses and data, will prove immensely beneficial to professionals and students involved in this area at various levels.

A foreword of all chapters of the book is provided below:

Chapter 1, Nuclear physics is a sub-discipline of physics that studies atomic nuclei, their constituents and their interactions. Its applications are in the fields of nuclear medicine, nuclear warfare, medical diagnostics, materials engineering, radiocarbon dating, power generation, etc. This is an introductory chapter which will discuss in brief about the scope and breadth of nuclear science and nuclear physics and elucidates its crucial principles; **Chapter 2**, An unstable atomic nucleus loses energy by emitting an alpha, beta particle, electron, etc. This process is termed as radioactivity. This chapter has been carefully written to provide an easy understanding of radioactivity through a detailed analysis of radionuclides, ionizing radiation, the law of radioactive decay, etc.; **Chapter 3**, An understanding of nuclear energetics follows from an in-depth study of nuclear energy, mass energy balance, binding energy, Q-value and nuclide stability. This chapter has been written to elucidate the fundamentals of these key concepts in a comprehensive language; **Chapter 4**, All the diverse interactions of particles with matter and the different forces acting on these have been covered in this chapter. The topics covered herein include interaction of proton with matter, interaction of neutron with matter, interaction of heavy charged particles with matter, weak nuclear force, strong nuclear force, etc.; **Chapter 5**, A radiation detector is an instrument that is used for tracking, detecting or identifying particles produced by nuclear decay, particle accelerator reactions or cosmic radiation. These devices can measure particle characteristics of energy, momentum, spin, charge, etc. This chapter strives to provide an overview of nuclear radiation detectors, particularly with reference to the bubble chamber, cloud chamber, gas detector, Cherenkov detectors, scintillation counters, etc.; **Chapter 6**, Nuclear fusion is the reaction characterized by the fusion of two or more atomic nuclei. In contrast, nuclear fission refers to the nuclear decay process in which an atomic nucleus splits into smaller elements, along with the release of a massive amount of energy. The topics discussed in this chapter include thermonuclear fusion, controlled fusion, fusion reaction, Maxwell-averaged fusion reactivities, etc. which will provide a thorough understanding of nuclear reactions.

At the end, I would like to thank all the people associated with this book devoting their precious time and providing their valuable contributions to this book. I would also like to express my gratitude to my fellow colleagues who encouraged me throughout the process.

Leonel Ware

1

Introduction to Nuclear Physics

Nuclear physics is a sub-discipline of physics that studies atomic nuclei, their constituents and their interactions. Its applications are in the fields of nuclear medicine, nuclear warfare, medical diagnostics, materials engineering, radiocarbon dating, power generation, etc. This is an introductory chapter which will discuss in brief about the scope and breadth of nuclear science and nuclear physics and elucidates its crucial principles.

Nuclear Science

Nuclear science is the study of sub-atomic particles and their application in various disciplines. It is crucial to understand our universe, our world, and ourselves at the atomic level. If we can understand how atoms come together, interact, and can be best combined with other atoms, then new, more efficient materials and medicines can be developed.

Nuclear Physics

Nuclear physics is a field of physics that involves investigation of the building blocks and interactions of atomic nuclei. It includes studies of nuclear components such as protons and neutrons, forces such as the strong force (or strong interaction), and phenomena such as radioactive decay, nuclear fission, and nuclear fusion.

Nuclear power and nuclear weapons are the most commonly known applications of nuclear physics, but the research field is also the basis for a far wider range of less common applications, such as in medicine (nuclear medicine, magnetic resonance imaging), materials engineering (ion implantation), and archaeology (radiocarbon dating).

Nuclear physics is sometimes used synonymously with atomic physics, but physicists usually differentiate between the two. Atomic physics studies the combined system of the atomic nucleus and the arrangement of electrons around the nucleus.

Particle physics involves study of the elementary constituents of matter and radiation, and the interactions between them. Particle physics evolved out of nuclear physics and, for this reason, has been included under the same term in earlier times.

Modern Nuclear Physics

A heavy nucleus can contain hundreds of nucleons, which means that with some approximation it can be treated as a classical system, rather than a quantum-mechanical one. In the resulting liquid-drop model, the nucleus has an energy that arises partly from surface tension and partly from electrical repulsion of the protons. The liquid-drop model is able to reproduce many features of nuclei, including the general trend of binding energy with respect to mass number, as well as the phenomenon of nuclear fission.

Superimposed on this classical picture, however, are quantum mechanical effects, which can be described using the nuclear shell model, developed in large part by Maria Goeppert-Mayer. Nuclei with certain numbers of neutrons and protons (the magic numbers 2, 8, 20, 50, 82, 126, ...) are particularly stable, because their shells are filled.

Other, more complicated models for the nucleus have also been proposed, such as the interacting boson model, in which pairs of neutrons and protons interact as bosons, analogously to Cooper pairs of electrons.

Much of current research in nuclear physics relates to the study of nuclei under extreme conditions, such as high spin and excitation energy. Nuclei may also have extreme shapes (similar to that of American footballs) or extreme neutron-to-proton ratios. Experimenters can create such nuclei using artificially induced fusion or nucleon transfer reactions, employing ion beams from a particle accelerator. Beams with even higher energies can be used to create nuclei at very high temperatures, and there are signs that these experiments have produced a phase transition from normal nuclear matter to a new state, the quark-gluon plasma, in which quarks mingle with one another, rather than being segregated in triplets as they are in neutrons and protons.

Nuclear Decay: Spontaneous Changes from one Nuclide to Another

There are 80 elements that have at least one stable isotope, and 250 such stable isotopes. However, there are thousands more well-characterized isotopes that are unstable. These radioisotopes may be unstable and decay along timescales ranging from fractions of a second to weeks, years, or even many millions of years.

If a nucleus has too few or too many neutrons it may be unstable, and will decay after some period of time. For example, in a process called beta decay, a nitrogen-16 atom (7 protons, 9 neutrons) is converted to an oxygen-16 atom (8 protons, 8 neutrons) within a few seconds of being created. In this decay, a neutron in the nitrogen nucleus is turned into a proton and an electron and antineutrino, by the weak nuclear force. The element is transmuted to another element in the process because, while it previously had seven protons (which makes it nitrogen), it now has eight (which makes it oxygen).

In alpha decay, the radioactive element decays by emitting a helium nucleus (2 protons and 2 neutrons), giving another element, plus helium-4. In many cases this process continues through several steps of this kind, including other types of decays, until a stable element is formed.

In gamma decay, a nucleus decays from an excited state into a lower energy state by emitting a gamma ray. The element is not changed in the process.

Other, more exotic decays, are also possible. For example, in internal conversion decay, the energy from an excited nucleus may be used to eject one of the inner orbital electrons from the atom. This process produces high speed electrons, but it is not beta decay, and (unlike beta decay) does not transmute one element to another.

Nuclear Fusion

When two light nuclei come into very close contact with each other, it is possible for the strong force to fuse the two together. It takes a great deal of energy to push the nuclei close enough together for the strong or nuclear forces to have an effect, so the process of nuclear fusion can take place only at very high temperatures or high densities. Once the nuclei are close enough together, the strong force overcomes their electromagnetic repulsion and squishes them into a new nucleus. A very large amount of energy is released when light nuclei fuse together because the binding energy per nucleon increases with mass number up until nickel-62.

Stars like the Sun are powered by the fusion of four protons into a helium nucleus, two positrons, and two neutrinos. The uncontrolled fusion of hydrogen into helium is known as thermonuclear runaway. Research to find an economically viable method of using energy from a controlled fusion reaction is currently being undertaken by various research establishments.

Nuclear Fission

For nuclei heavier than nickel-62, the binding energy per nucleon decreases with mass number. It is therefore possible for energy to be released if a heavy nucleus breaks apart into two lighter ones. This splitting of atomic nuclei is known as nuclear fission.

The process of alpha decay may be thought of as a special type of spontaneous nuclear fission. This process produces a highly asymmetrical fission because the four particles that make up the alpha particle are especially tightly bound to each other, making production of this nucleus in fission particularly likely.

For some of the heaviest nuclei that produce neutrons on fission, and which also easily absorb neutrons to initiate fission, a self-igniting type of neutron-initiated fission can be obtained, in a so-called chain reaction.

The fission or "nuclear" chain-reaction, using fission-produced neutrons, is the source of energy for nuclear power plants and fission type nuclear bombs such as the two that the United States used against Hiroshima and Nagasaki to bring an end to World War II in the Pacific theater. Heavy nuclei such as uranium and thorium may undergo spontaneous fission, but they are much more likely to undergo alpha decay.

For a neutron-initiated chain-reaction to occur, there must be a critical mass of the element present in a certain space under certain conditions (these conditions slow and conserve neutrons for the reactions). There is one known example of a natural nuclear fission reactor, which was active in two regions of Oklo, Gabon, Africa, over 1.5 billion years ago. Measurements of natural neutrino emission have demonstrated that around half of the heat emanating from the Earth's core results from radioactive decay. However, it is not known if any of this results from fission chain-reactions.

Production of Heavy Elements

As the Universe cooled after the Big Bang, it eventually became possible for particles as we know them to exist. The most common particles created in the Big Bang that are still easily observable to us were protons (hydrogen) and electrons (in equal numbers). Some heavier elements were created as the protons collided with each other, but most of the heavy elements we see today were created inside of stars during a series of fusion stages, such as the proton-proton chain, the CNO cycle, and the triple-alpha process.

Progressively heavier elements are created during the evolution of a star. Since the binding energy per nucleon peaks around iron, energy is released only through fusion processes occurring below this point. Since the creation of heavier nuclei by fusion costs energy, nature resorts to the process of neutron capture. Neutrons (due to their lack of charge) are readily absorbed by a nucleus. The heavy elements are created by either a slow neutron capture process (the so-called s process) or by the rapid (or r) process. The s process occurs in thermally pulsing stars (called AGB, or asymptotic giant branch stars) and takes hundreds to thousands of years to reach the heaviest elements of lead and bismuth. The r process is thought to occur in supernova explosions because the conditions of high temperature, high neutron flux and ejected matter are present. These stellar conditions make the successive neutron captures very fast, involving very neutron-rich species, which then undergo beta decay to heavier elements, especially at the so-called waiting points that correspond to more stable nuclides with closed neutron shells (magic numbers). The r process duration is typically in the range of a few seconds.

2

Radioactivity

An unstable atomic nucleus loses energy by emitting an alpha, beta particle, electron, etc. This process is termed as radioactivity. This chapter has been carefully written to provide an easy understanding of radioactivity through a detailed analysis of radionuclides, ionizing radiation, the law of radioactive decay, etc.

The property of emission of radioactive rays from radioactive elements is termed as radioactivity.

Generally, elements with atomic number more than 82 show radioactivity and disintegrated to small nuclei with the emission of alpha, beta, proton, neutron particles or gamma rays. This nuclei with decomposed is called as parent nuclei and the product nuclei is termed as daughter nuclei.

The atomic number and mass depends upon the type of radioactive rays emitted during nuclear reaction. The decay of radioactive parent nuclei to stable nuclei is known as radioactive decay or nuclear decay.

The type of decay depends on the type of radioactive particles emitted in decay. For example,

1. Alpha decay
2. Beta decay
3. Gamma decay

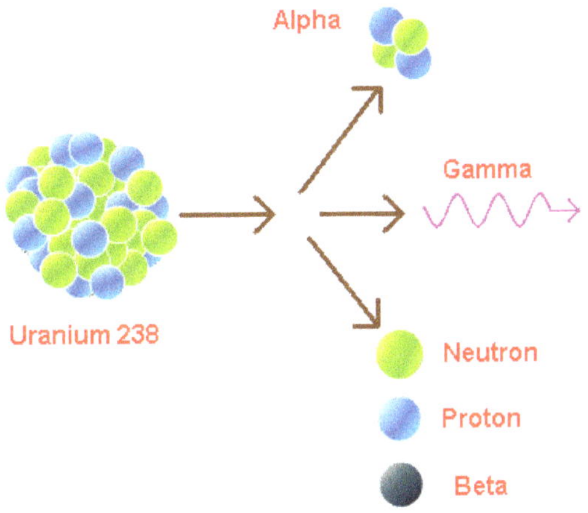

Radioactivity Units

The unit of radioactivity called as curie (ci). One curie defined as "the amount of a radioactive substance which has a decay rate of 3.7 x 1010 disintegration's per second." This large number is based on the observation that one gram of radium disintegration at the rate of 3.7 x 1010 disintegration per second.

The non SI unit of radioactivity is defined as curie which is equals to amount of a radioactive substance which has a decay rate of 3.7×10^{10} disintegration per second given after the study of radium isotope ^{226}Ra by the Curies.

$$1\,Ci = 3.7 \times 10^{10} \text{ decays / second}$$

The SI unit of radioactivity is called as becquerel (Bq), which is equals to one decay per second.

$$1 \text{ becquerel} = 1 \text{ radioactive decay / second} = 2.703 \times 10^{-11} \text{ Curie}$$

$$1 \text{ Curie} = 3.7 \times 10^{10} \text{ Becquerel} = 37 \text{ Giga Becquerel}$$

$$1\,Bq \cong 2.703 \times 10^{-11}\,Ci$$

For small measurements; millicurie (mci) and microcurie(μci) is used.

$$1\,mci = 3.7 \times 10^{7} \text{ disintegration/ sec}$$

$$1\,\mu ci = 3.7 \times 10^{4} \text{ disintegration/ sec or } 2.22 \times 10^{6} \text{ disintegrations per minute}$$

The relation between curie and becquerel (Bq) with decay constant is as follows.

$$N(\text{atoms}) * \lambda(1/s) = 1\,Ci = 3.7 \times 10^{10}(Bq)$$

Where,

λ = Decay constant in (1/s)

N = Amount of radioactive substance

Types of Radioactivity

The unstable radioactive elements show nuclear decay which is a spontaneous cleavage of an atomic nucleus with the formation of some other nucleus and release a certain amount of energy. Since, radioisotopes have unstable nuclei which do not have enough binding energy to hold the nucleus together and constantly try to get stabilize.

This process of transformation of one unstable nucleus to stable nuclei by the emission of energy in the form of radiation is called as transmutation. The nuclear decay and transmutation process will continue until a new formed element also called as daughter nuclei has a stable nucleus and not radioactive in nature.

The radioactive decay or transmutation can occur naturally or by artificial means. On this basis, transmutation classified as two types.

- Natural radioactivity
- Artificial radioactivity

Natural Radioactivity

The atoms of radioactive elements on the emission of alpha particles and beta particles would change into atoms of another element. This change is spontaneous and occurs due to instability of heavy nuclei. Such type of radioactive decay termed as natural radioactive decay and phenomenon called as natural radioactivity.

Rutherford was first to observed the decay of radon by the loss of alpha and beta particles by the preceding elements. Natural radioactive elements decay naturally without any external effect until the convert in stable nuclei.

For example; Uranium-238 disintegrated in to Thorium-234 by the emission of alpha particle which further changes in Protactinium-234 by the loss of a beta particle and anti-neutrino.

$$92U^{238} \rightarrow 90TH^{234} + 2He^{4}$$
$$90Th^{234} \rightarrow 91Pa^{234} + {}_{-1}e^{0} + \ddot{I}...$$

The emission of an alpha particle results in the formation of an element which lie two place to the left in periodic table and the emission of beta particle results in the formation of an element which lies one place to the right. This is called as group displacement law. Natural radioactivity series.

The uranium series starts from uranium-23 and finally converts to lead-206 by the emission of eight alpha particles and six beta particles.

$$92^{U^{238}} \rightarrow 82^{Pb^{206}} + 8_2He^4 + 6_{-1}e^0$$

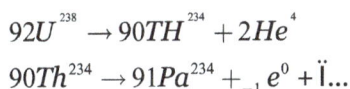

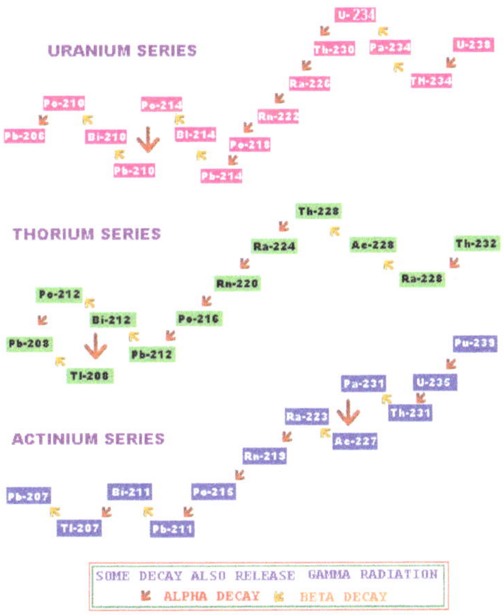

- All first three series; Uranium, Thorium and actinium series end with an isotope of lead that is Pb-206, Pb-208, Pb-207.
- The mass number of all elements in thorium series are multiples of 4, hence can be represented as 4n series.

- In the same way uranium series and actinium series can be represented as 4n+2 and 4n+3 series respectively.
- The neptunium series end with an isotope of bismuth-209.
- The neptunium series represented as 4n+1 series.

Artificial Radioactivity

Sometimes radioactivity created artificially in elements that are called as artificial radioactivity. Artificial radioactivity can be created in two ways.

- Artificial transmutation
- Induced transmutation

Artificial Transmutation

The conversion of one element into another by artificial means is known as artificial transmutation. It first observed in 1919 by Rutherford during the bombardment of alpha particles on the nucleus of nitrogen to form oxygen isotope and proton.

$$_7N^{14} + _2He^4 \rightarrow _8O^{17} + _1H^1$$

Later in 1932 the discovery of neutrons by James Chadwick was also an application of artificial transmutation in which he used the bombardment of alpha particles on the nucleus of beryllium to form neutron and carbon nucleus.

$$_4Be^9 + _2He^4 \rightarrow _6C^{12} + _0n^1$$

Artificial transmutation can be done by using several particles like; alpha particles, neutrons , deuteron and protons. For example,

$$_4Be^0 + _2He^4 \rightarrow _6C^{12} + _0n^1$$

$$_{94}Pu^{239} + _2He^4 \rightarrow _{94}Cm^{242} + _0n^1$$

$$_9F^{19} + _2He^4 \rightarrow _{10}Ne^{22} + _1H^1$$

$$_7N^{14} + _2He^4 \rightarrow _8O^{17} + _1H^1$$

$$_{26}Fe^{59} + _2He^4 \rightarrow _{29}Cu^{63} + _{-1}e^0$$

$$_{15}P^{31} + _1H^1 \rightarrow _{16}S^{31} + _0n^1$$

$$_6C^{12} + _1H^1 \rightarrow N^{13} + \gamma$$

$$_4Be^9 + _1H^1 \rightarrow _4Be^8 + _1H^2$$

$${}_{8}O^{16} + {}_{1}H^{1} \rightarrow {}_{7}N^{13} + {}_{2}He^{4}$$

$${}_{13}Al^{27} + {}_{0}n^{1} \rightarrow {}_{12}Mg^{27} + {}_{1}H^{1}$$

$${}_{8}O^{16} + {}_{0}n^{1} \rightarrow {}_{6}C^{13} + {}_{2}He^{4}$$

$${}_{92}U^{238} + {}_{0}n^{1} \rightarrow {}_{92}U^{238} + \lambda$$

$${}_{8}O^{18} + {}_{0}n^{1} \rightarrow {}_{9}F^{19} + {}_{-1}e^{0}$$

$${}_{3}Li^{6} + {}_{1}H^{2} \rightarrow 3Li7 + {}_{1}H^{1}$$

$${}_{32}As^{75} + {}_{1}H^{2} \rightarrow {}_{32}As^{76} + {}_{1}H^{1}$$

$${}_{4}Be^{9} + \gamma \rightarrow {}_{4}Be^{8} + {}_{0}n^{1}$$

Uses of Radioactivity

The radioactive elements used in many different fields. Like; in atomic energy, agriculture, in different industries and in rock dating process.

- Discovery of new elements particles: The study of radioactivity has led to the discovery of new fundamental particles like neutrons, positrons, deuterons, alpha particles etc. These particles are highly useful in causing artificial transmutation of elements and adding to our knowledge of atomic structure.
- Discovery of isotopes and isobars: The isotopes and isobars were first discovered in radioactive series and were afterwards looked for amongst non-radioactive elements.
- Discovery of new elements: The discovery of trans uranic elements of atomic numbers higher than 92 not known to exist in nature. A large number of different radioactive isotopes of known elements have been discovered by artificial means.
- Release of atomic energy: Nuclear reactions like nuclear fusion and nuclear fission release enormous amount of energy which can be used for different purpose.
- Radioactive tracer: Many radioactive isotopes are used in tracing various processes in surgery, medicine, biology, agriculture, industry and chemistry. In tracer technique a radioactive isotope or its compound is introduced at one point of a system and its movement is then trace by measuring radioactivity in different parts of the system. Such isotopes are known as radioactive tracers. For example; Phosphate containing radioactive isotope of phosphorus is used for patients suffering from bone fracture to check of the phosphorus is being absorbed by the bone or not. Similarly the injection of radioactive iodine Uptake is a test of thyroid function.
- Rock dating and carbon dating: Radioactive isotope of carbon(C-14) is used to estimate the age of earth and for the estimation of age of fossils. The half life for C-14 is around 5568 years.

$$6^{C^{14}} \rightarrow 7^{N^{14}+beta\, particles}$$

$$7^{N^{14}} + 0^{n^1} \rightarrow 6^{C^{14}} + 1^{H^1}$$

Because of these two reactions, the quantity of C-14 and carbon dioxide $^{12}CO_2$ present in the atmosphere has been constant over long periods of years This C-14 has been constant over long periods of years. This C-14 has been consumed by plants as well as by other living organism and remains constant for a long time, hence can be detected easily.

Effects of Radioactivity

With all these advantage of radioactivity; the over dose of radioactive substances show so many adverse effect like; nausea, vomiting, headache and some loss of blood cells. Some adverse effects of radioactive rays are as follows.

- The radiation exposure at 200 rems or higher creates clumps and responsible for quick hair loss.
- Brain cell damages only at a very high exposure of radioactive rays around 5,000 rems or greater.
- Some parts of body are more susceptible for exposure to radioactive radiation sources. For example; thyroid gland is susceptible for radioactive iodine, hence radioactive iodine can destroy thyroid gland.
- The exposure of radioactive radiations around 100 rems affected the blood system by decreasing the count of the blood's lymphocyte cell which makes the victim more susceptible to infection known as mild radiation sickness.
- Intense exposure to radioactive substances can damage to small blood vessels and cause heart failure and death directly.
- Radioactive rays can damage the intestinal tract and cause bloody vomiting, nausea and diarrhea.

The adverse effect of radioactive radiations had been seen in real life events like,

- Hiroshima and Nagasaki nuclear accident occur in 1945 during World War II.
- Three Mile Island accident happen in nuclear power plant accidents on 28th march 1979.
- Chernobyl nuclear accident occurred at the Chernobyl Nuclear Power Plant in Ukraine on 26 April 1986.

Radioactivity Detector

The radioactivity detector is used for the detection of radioactive rays like; alpha, beta and gamma rays. The Geiger counter, or Geiger-Muller counter, is a common radioactive detector which used to measures ionizing radiation.

In Geiger-Muller counter detects the radiation produced by radioactive substance at low-pressure gas filled in a Geiger-Muller tube. An inert gas filled Geiger-Muller tube which becomes conductive due to radiation and conduct electricity. This electricity displayed by a needle or lamp and/or audible clicks.

Geiger Muller counter invented in 1908 but still a popular instruments which can be used for measurements in health, industry, physics, geology and other fields due to the capability of making simple electronic circuits. Various form of Geiger-Muller detectors are available like GM-40, GM-10 widely used in different fields.

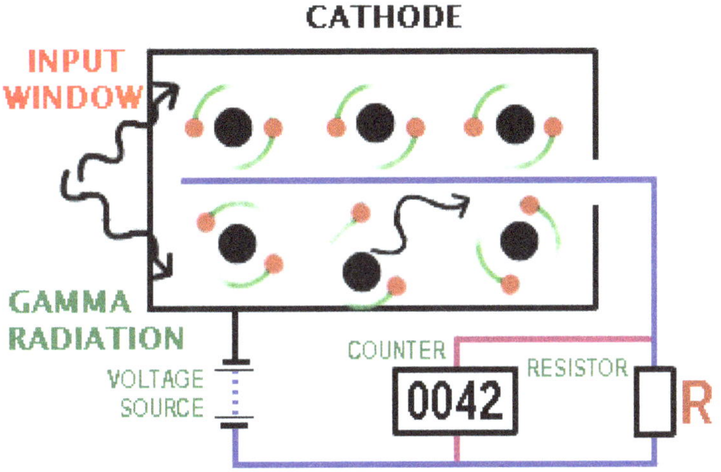

Radionuclide

A radioactive nuclide radionuclides or radionuclide is an unstable nuclide and thus degenerates emitting ionizing radiation. Although some physicists sometimes commonly used to designate the palabra radioisótopo however, it should be noted that strict or formal language of physics and nuclear technology is flawed as a nuclide and an isotope are not the same. When a radionuclide emits Radioctividad alcanza a more stable state, which requires less energy than before, and generally becomes a different nuclide (or in the same, but less excited, if issued gamma radioactivity), which may or not was also radioactive. This radioactive process occurs spontaneously in principle, but man has learned to cause artificially. In either case the resulting radioactivity has exactly the same characteristics.

Radionuclides are characterized by a finite lifetime, which can range from tiny fractions of a second to thousands of years. In fact, some of them have such long half-life that has not yet been quantified experimentally and even some that were considered, and for some practical applications but is usually made stable. There currently known nuclides ninety theoretically stable and two hundred fifty-five those who were not observed disintegrate. Moreover, there are nearly twice, about six hundred fifty, which if they have been observed radioactivity and have a half life of at least one hour. On Earth some three thousand radionuclides than one hour, most of which are known half-life (about 90%) are anthropogenic (caused by humans), a 2400 lower one hour half life and still others so unstable that its half-life is very short.

Radionuclides apply to nuclear technology for electricity in industry (quality control, etc.), medicine (radiation, etc.) and nuclear weapons (primarily for the propulsion of vehicles and tools to kill). Its use implies serious environmental risks (radioactive contamination) and health (radio toxicity, radiation poisoning, etc.), so it must be done with extreme care. Recall that naturally occurring radionuclides such as uranium or plutonium exist in finite quantities on Earth, so you have to use them in a sustainable manner. Moreover, their use generates radioactive waste, which can be very dangerous and for which the only treatment is usually done is cover until its radioactivity is close to the natural.

Ionizing Radiation

Ionizing radiation, also (imprecisely) called radioactivity, is electromagnetic (EM) radiation whose waves contain energy sufficient to overcome the binding energy of electrons in atom s or molecules, thus creating ions. The wavelength is shorter than that of ultraviolet (UV).

Ionizing radiation can occur as a barrage of photon s having a nature similar to that of visible light, but with far shorter wavelength and consequently higher frequency. This type of radiation includes X rays and gamma rays. More massive particles also comprise ionizing radiation if they travel at sufficient speed. These include high-speed electrons (beta particles), protons, neutrons, and helium nuclei (alpha particles). Ionizing radiation is dangerous because it damages the internal structures of living cells. This can cause cell death in high doses over a short period of time, and errors in the reproductive process (mutations) in lower doses over longer periods of time.

Examples of non-ionizing EM radiation include radio (RF) waves, extremely low frequency (ELF) fields, infrared (IR), visible light, and UV. These forms of EM energy are generally not dangerous, with some exceptions: high-energy radio microwave s and IR which can cause destructive heating of biological tissue; intense visible light which can cause blindness; and intense UV which can cause blindness and superficial skin burns in high doses over a short period of time, and skin cancer and cataracts of the eye at lower doses over long periods of time. There is debate as to whether long-term exposure to moderate-to-intense radio-frequency (RF) fields and ELF fields is harmful to human beings.

The most common unit of ionizing radiation is the becquerel (Bq), equal to one disintegration or nuclear transformation per second. Reduced to base units in the International System of Units (SI $1\,\mathrm{Bq} = 1/\mathrm{s}$ or $1\,\mathrm{s}^{-1}$. An alternative unit is the curie (Ci), equivalent to 3.7×10^{10} disintegrations per second or 2.2×10^{12} disintegrations per minute. To convert from curies to becquerels, multiply by 3.7×10^{10}. To convert from becquerels to curies, multiply by 2.7×10^{-11}.

Types of Ionizing Radiation

There are many types of ionizing radiation, but the most familiar are alpha, beta, and gamma/x-ray radiation. Neutrons, when expelled from atomic nuclei and traveling as a form of radiation, can also be a significant health concern.

Alpha particles are clusters of two neutrons and two protons each. They are identical to the nuclei of atoms of helium, the second lightest and second most common element in the universe, after hydrogen. Compared with other forms of radiation, though, these are very heavy particles--about 7,300 times the mass of an electron. As they travel along, these large and heavy particles frequently interact with the electrons of atoms, rapidly losing their energy. They cannot even penetrate a piece of paper or the layer of dead cells at the surface of our skin. But if released within the body from a radioactive atom inside or near a cell, alpha particles can do great damage as they ionize atoms, disrupting living cells. Radium and plutonium are two examples of alpha emitters.

Beta particles are electrons traveling at very high energies. If alpha particles can be thought of as large and slow bowling balls, beta particles can be visualized as golf balls on the driving range. They travel farther than alpha particles and, depending on their energy, may do as much damage. For example, beta particles in fallout can cause severe burns to the skin, known as beta burns. Radiosotopes that emit beta particles are present in fission products produced in nuclear reactors and nuclear explosions. Some beta-emitting radioisotopes, such as iodine 131, are administered internally to patients to diagnose and treat disease.

Gamma and x-ray radiation consists of packets of energy known as photons. Photons have no mass or charge, and they travel in straight lines. The visible light seen by our eyes is also made up of photons, but at lower energies. The energy of a gamma ray is typically greater than 100 kiloelectron volts (keV-"k" is the abbreviation for kilo, a prefix that multiplies a basic unit by 1,000) per photon, more than 200,000 times the energy of visible light (0.5 eV). If alpha particles are visualized as bowling balls and beta particles as golf balls, photons of gamma and x-radiation are like weightless bullets moving at the speed of light. Photons are classified according to their origin. Gamma rays originate from events within an atomic nucleus; their energy and rate of production depend on the radioactive decay process of the radionuclide that is their source. X rays are photons that usually originate from energy transitions of the electrons of an atom. These can be artificially generated by bombarding appropriate atoms with high-energy electrons, as in the classic x-ray tube. Because x rays are produced artificially by a stream of electrons, their rate of output and energy can be controlled by adjusting the energy and amount of the electrons themselves. Both x rays and gamma rays can penetrate deeply into the human body. How deeply they penetrate depends on their energy; higher energy results in deeper penetration into the body. A 1 MeV ("M" is the abbreviation for mega, a prefix that multiplies a basic unit by 1,000,000) gamma ray, with an energy 2,000,000 times that of visible light, can pass completely through the body, creating tens of thousands of ions as it does.

A final form of radiation of concern is neutron radiation. Neutrons, along with protons, are one of the components of the atomic nucleus. Like protons, they have a large mass; unlike protons, they have no electric charge, allowing them to slip more easily between atoms. Like a Stealth fighter, high-energy neutrons can travel farther into the body, past the protective outer layer of the skin, before delivering their energy and causing ionization.

Indirectly Ionizing

Indirect ionizing radiation is electrically neutral and therefore does not interact strongly with matter. The bulk of the ionization effects are due to secondary ionizations.

An example of indirectly ionizing radiation is neutron radiation.

Photon Radiation

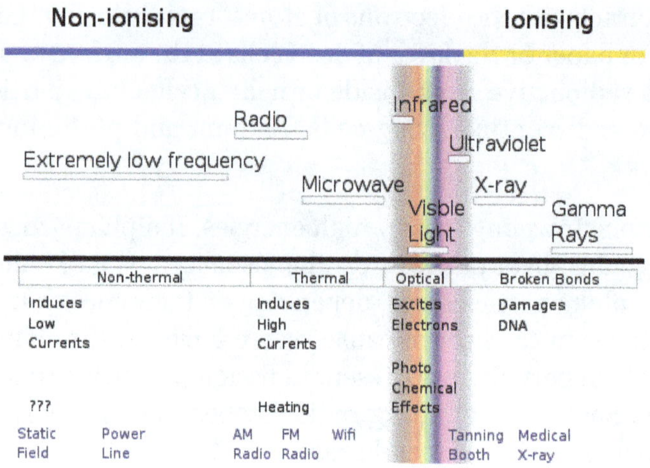

Different types of electromagnetic radiation

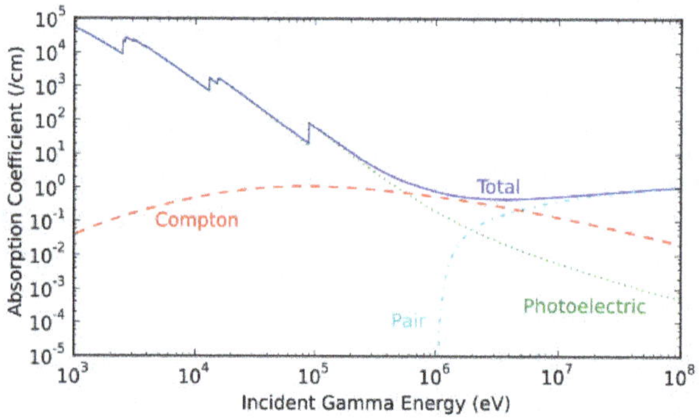

The total absorption coefficient of lead (atomic number 82) for gamma rays, plotted versus gamma energy, and the contributions by the three effects. Here, the photoelectric effect dominates at low energy. Above 5 MeV, pair production starts to dominate.

Even though photons are electrically neutral, they can ionize atoms directly through the photoelectric effect and the Compton effect. Either of those interactions will cause the ejection of an electron from an atom at relativistic speeds, turning that electron into a beta particle (secondary beta particle) that will ionize many other atoms. Since most of the affected atoms are ionized directly by the secondary beta particles, photons are called indirectly ionizing radiation.

Photon radiation is called gamma rays if produced by a nuclear reaction, subatomic particle decay, or radioactive decay within the nucleus. It is otherwise called x-rays if produced outside the nucleus. The generic term photon is therefore used to describe both.

X-rays normally have a lower energy than gamma rays, and an older convention was to define the boundary as a wavelength of 10^{-11} m or a photon energy of 100 keV. That threshold was driven by limitations of older X-ray tubes and low awareness of isomeric transitions. Modern technologies and discoveries have resulted in an overlap between X-ray and gamma energies. In many fields they are functionally identical, differing for terrestrial studies only in origin of the radiation. In astronomy, however, where radiation origin often cannot be reliably determined, the old energy division has been preserved, with X-rays defined as being between about 120 eV and 120 keV, and gamma rays as being of any energy above 100 to 120 keV, regardless of source. Most astronomical "gamma-ray astronomy" are known *not* to originate in nuclear radioactive processes but, rather, result from processes like those that produce astronomical X-rays, except driven by much more energetic electrons.

Photoelectric absorption is the dominant mechanism in organic materials for photon energies below 100 keV, typical of classical X-ray tube originated X-rays. At energies beyond 100 keV, photons ionize matter increasingly through the Compton effect, and then indirectly through pair production at energies beyond 5 MeV. The accompanying interaction diagram shows two Compton scatterings happening sequentially. In every scattering event, the gamma ray transfers energy to an electron, and it continues on its path in a different direction and with reduced energy.

Definition Boundary for Lower-energy Photons

The lowest ionization energy of any element is 3.89 eV, for caesium. However, US Federal Communications Commission material defines ionizing radiation as that with a photon energy greater than 10 eV (equivalent to a far ultraviolet wavelength of 124 nanometers). Roughly, this corresponds to both the first ionization energy of oxygen, and the ionization energy of hydrogen, both about 14 eV. In some Environmental Protection Agency references, the ionization of a typical water molecule at an energy of 33 eV is referenced as the appropriate biological threshold for ionizing radiation: this value represents the so-called *W-value*, the colloquial name for the ICRU's mean energy expended in a gas per ion pair formed, which combines ionization energy plus the energy lost to other processes such as excitation. At 38 nanometers wavelength for electromagnetic radiation, 33 eV is close to the energy at the conventional 10 nm wavelength transition between extreme ultraviolet and X-ray radiation, which occurs at about 125 eV. Thus, X-ray radiation is always ionizing, but only extreme-ultraviolet radiation can be considered ionizing under all definitions.

As noted, the biological effect of ionizing radiation on cells somewhat resembles that of a broader spectrum of molecularly damaging radiation, which overlaps ionizing radiation and extends beyond, to somewhat lower energies into all regions of UV and sometimes visible light in some systems (such as photosynthetic systems in leaves). Although DNA is always susceptible to damage by ionizing radiation, the DNA molecule may also be damaged by radiation with enough energy to excite certain molecular bonds to form thymine dimers. This energy may be less than ionizing, but near to it. A good example is ultraviolet spectrum energy which begins at about 3.1 eV (400 nm) at close to the same energy level which can cause sunburn to unprotected skin, as a result of photoreactions in collagen and (in the UV-B range) also damage in DNA (for example, pyrimidine dimers). Thus, the mid and lower ultraviolet electromagnetic spectrum is damaging to biological tissues as a result of electronic excitation in molecules which falls short of ionization, but produces similar non-thermal effects. To some extent, visible light and also ultraviolet A (UVA) which is closest to visible energies, have been proven to result in formation of reactive oxygen species in skin, which cause indirect damage since these are electronically excited molecules which can inflict reactive damage, although they do not cause sunburn (erythema). Like ionization-damage, all these effects in skin are beyond those produced by simple thermal effects.

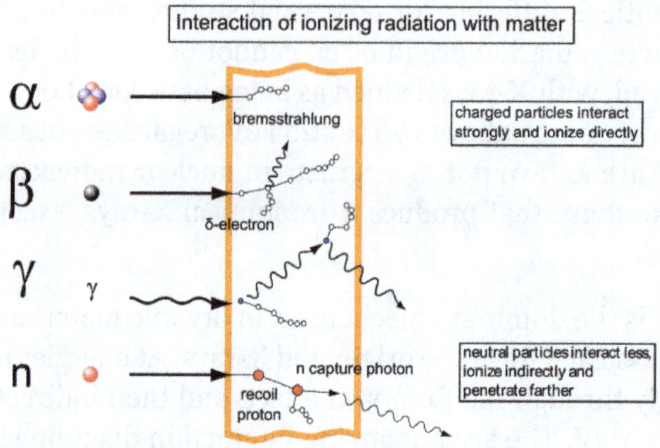

Radiation interaction: gamma rays are represented by wavy lines, charged particles and neutrons by straight lines. The small circles show where ionization occurs.

Neutrons

Neutrons have zero electrical charge and thus often do not *directly* cause ionization in a single step or interaction with matter. However, fast neutrons will interact with the protons in hydrogen via LET, and this mechanism scatters the nuclei of the materials in the target area, causing direct ionization of the hydrogen atoms. When neutrons strike the hydrogen nuclei, proton radiation (fast protons) results. These protons are themselves ionizing because they are of high energy, are charged, and interact with the electrons in matter.

Neutrons that strike other nuclei besides hydrogen will transfer less energy to the other particle if LET does occur. But, for many nuclei struck by neutrons, inelastic scattering occurs. Whether elastic or inelastic scatter occurs is dependent on the speed of the neutron, whether fast or thermal or somewhere in between. It is also dependent on the nuclei it strikes and its neutron cross section.

In inelastic scattering, neutrons are readily absorbed in a process called neutron capture and attributes to the neutron activation of the nucleus. Neutron interactions with most types of matter in this manner usually produce radioactive nuclei. The abundant oxygen-16 nucleus, for example, undergoes neutron activation, rapidly decays by a proton emission forming nitrogen-16, which decays to oxygen-16. The short-lived nitrogen-16 decay emits a powerful beta ray. This process can be written as:

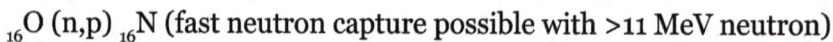

${}_{16}O$ (n,p) ${}_{16}N$ (fast neutron capture possible with >11 MeV neutron)

${}_{16}N \rightarrow {}_{16}O + \beta^-$ (Decay $t_{1/2} = 7.13$ s)

This high-energy β^- further interacts rapidly with other nuclei, emitting high-energy γ via Bremsstrahlung

While not a favorable reaction, the ${}_{16}O$ (n,p) ${}_{16}N$ reaction is a major source of X-rays emitted from the cooling water of a pressurized water reactor and contributes enormously to the radiation generated by a water-cooled nuclear reactor while operating.

For the best shielding of neutrons, hydrocarbons that have an abundance of hydrogen are used.

In fissile materials, secondary neutrons may produce nuclear chain reactions, causing a larger amount of ionization from the daughter products of fission.

Outside the nucleus, free neutrons are unstable and have a mean lifetime of 14 minutes, 42 seconds. Free neutrons decay by emission of an electron and an electron antineutrino to become a proton, a process known as beta decay:

In the above diagram, a neutron collides with a proton of the target material, and then becomes a fast recoil proton that ionizes in turn. At the end of its path, the neutron is captured by a nucleus in an (n,γ)-reaction that leads to the emission of a neutron capture photon. Such photons always have enough energy to qualify as ionizing radiation.

Physical Effects

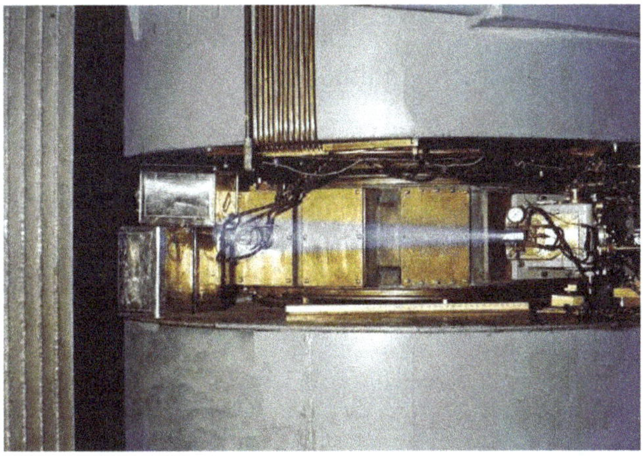

Ionized air glows blue around a beam of particulate ionizing radiation from a cyclotron

Nuclear Effects

Neutron radiation, alpha radiation, and extremely energetic gamma (> ~20 MeV) can cause nuclear transmutation and induced radioactivity. The relevant mechanisms are neutron activation, alpha absorption, and photodisintegration. A large enough number of transmutations can change macroscopic properties and cause targets to become radioactive themselves, even after the original source is removed.

Chemical Effects

Ionization of molecules can lead to radiolysis (breaking chemical bonds), and formation of highly reactive free radicals. These free radicals may then react chemically with neighbouring materials even after the original radiation has stopped. (e.g., ozone cracking of polymers by ozone formed by ionization of air). Ionizing radiation can disrupt crystal lattices in metals, causing them to become amorphous, with consequent swelling, material creep, and embrittlement. Ionizing radiation can also accelerate existing chemical reactions such as polymerization and corrosion, by contributing to the activation energy required for the reaction. Optical materials darken under the effect of ionizing radiation.

High-intensity ionizing radiation in air can produce a visible ionized air glow of telltale bluish-purplish color. The glow can be observed, e.g., during criticality accidents, around mushroom clouds shortly after a nuclear explosion, or inside of a damaged nuclear reactor like during the Chernobyl disaster.

Monatomic fluids, e.g. molten sodium, have no chemical bonds to break and no crystal lattice to disturb, so they are immune to the chemical effects of ionizing radiation. Simple diatomic compounds with very negative enthalpy of formation, such as hydrogen fluoride will reform rapidly and spontaneously after ionization.

Electrical Effects

Ionization of materials temporarily increases their conductivity, potentially permitting damaging current levels. This is a particular hazard in semiconductor microelectronics employed in electronic equipment, with subsequent currents introducing operation errors or even permanently damaging the devices. Devices intended for high radiation environments such as the nuclear industry and extra atmospheric (space) applications may be made *radiation hard* to resist such effects through design, material selection, and fabrication methods.

Proton radiation found in space can also cause single-event upsets in digital circuits.

The electrical effects of ionizing radiation are exploited in gas-filled radiation detectors, e.g. the Geiger-Muller counter or the ion chamber.

Biological Effects of Ionizing Radiation

From the time that radioactivity was discovered, it was obvious that it caused damage. As early as 1901, Pierre Curie discovered that a sample of radium placed on his skin produced wounds that were very slow to heal. What some find surprising is the magnitude of the difference between the biological effects of non-ionizing radiation, such as light and microwaves, and ionizing radiation, such as high-energy ultraviolet radiation, $x - rays, gamma - rays,$ *and alpha – or* ◆/i > *–particles.*

Radiation at the low-energy end of the electromagnetic spectrum, such as radio waves and microwaves, excites the movement of atoms and molecules, which is equivalent to heating the sample. Radiation in or near the visible portion of the spectrum excites electrons into higher-energy orbitals. When the electron eventually falls back to a lower-energy state, the excess energy is given off to neighboring molecules in the form of heat. The principal effect of non-ionizing radiation is therefore an increase in the temperature of the system.

We experience the fact that biological systems are sensitive to heat each time we cook with a microwave oven, or spend too long in the sun. But it takes a great deal of non-ionizing radiation to reach dangerous levels. We can assume, for example, that absorption of enough radiation to produce an increase of about 6C in body temperature would be fatal. Since the average 70-kilogram human is 80% water by weight, we can use the heat capacity of water to calculate that it would take about 1.5 million joules of non-ionizing radiation to kill the average human. If this energy was carried by visible light with a frequency of $5\ x\ 10^{14}\ s^{-1}$,, it would correspond to absorption of about seven moles of photons.

Ionizing radiation is much more dangerous. A dose of only 300 joules of x-ray or gamma-ray radiation is fatal for the average human, even though this radiation raises the temperature of the body by only $0.001 C$. $\alpha - particle\ radiation$ is even more dangerous; a dose equivalent to only 15 joules is fatal for the average human. Whereas it takes seven moles of photons of visible light to produce a fatal dose of non-ionizing radiation, absorption of only 7×10^{-10} moles of the alpha-particles emitted by ^{238}U is fatal.

There are three ways of measuring ionizing radiation:

- Measure the activity of the source in units of disintegrations per second or curies, which is the easiest measurement to make.
- Measure the radiation to which an object is exposed in units of roentgens by measuring the amount of ionization produced when this radiation passes through a sample of air.
- Measure the radiation absorbed by the object in units of radiation absorbed doses or "rads." This is the most useful quantity, but it is the hardest to obtain.

One radiation absorbed dose, or rad, corresponds to the absorption of 10^{-5} joules of energy per gram of body weight. Because this is equivalent to $0.01\ J/kg$, one rad produces an increase in body temperature of about $2 \times 10^{-6} C$. At first glance, the rad may seem to be a negligibly small unit of measurement. The destructive power of the radicals produced when water is ionized is so large, however, that cells are inactivated at a dose of 100 rads, and a dose of 400 to 450 rads is fatal for the average human.

Not all forms of radiation have the same efficiency for damaging biological organisms. The faster energy is lost as the radiation passes through the tissue, the more damage it does. To correct for the differences in radiation biological effectiveness (RBE) among various forms of radiation, a second unit of absorbed dose has been defined. The roentgen equivalent man, or rem, is the absorbed dose in rads times the biological effectiveness of the radiation.

$$rems = rads \times RBE$$

Values for the RBE of different forms of radiation are given in the table below.

The Radiation Biological Effectiveness of Various Forms of Radiation

Radiation	RBE
$x - rays\ and\ gamma - rays$	1
$- particles\ with\ energies\ larger\ than\ 0.03\ MeV$	1
$- particles\ with\ energies\ less\ than\ 0.3\ MeV$	1.7
$thermal\ (slow - moving)\ neutrons$	3
$fast - moving\ neutrons\ or\ protons$	10
$\alpha - particles\ or\ heavy\ ions$	20

Estimates of the per capita exposure to radiation in the United States are summarized in the table below. These estimates include both external and internal sources of natural background radiation.

Average Whole-Body Exposure Levels for Sources of Ionizing Radiation

Source		Per Capita Dose $(rems\ /\ y)$
natural background		0.082
medical x-rays		0.077
nuclear test fallout		0.005
consumer and industrial products		0.005
nuclear power industry		0.001
	total:	0.170

External sources include cosmic rays from the sun and alpha-particles or $\lambda - rays$ emitted from rocks and soil. Internal sources include nuclides that enter the body when we breathe $({}^{14}C, {}^{85}Kr, {}^{220}Rn,\ and\ {}^{222}Rn)$ and through the food chain $({}^{3}H, {}^{14}C, {}^{40}K, {}^{90}Sr, {}^{131}I,\ and\ {}^{137}Cs)$. The actual dose from natural radiation depends on where one lives. People who live in the Rocky Mountains, for example, receive twice as much background radiation as the national average because there is less atmosphere to filter out the cosmic rays from the sun. The average dose from medical x-rays has decreased in recent years because of advances in the sensitivity of the photographic film used for x-rays. Radiation from nuclear test fallout has also decreased as a result of the atmospheric nuclear test ban.

Uses

Ionizing radiation has many industrial, military, and medical uses. Its usefulness must be balanced with its hazards, a compromise that has shifted over time. For example, at one time, assistants in shoe shops used X-rays to check a child's shoe size, but this practice was halted when the risks of ionizing radiation were better understood.

Neutron radiation is essential to the working of nuclear reactors and nuclear weapons. The penetrating power of x-ray, gamma, beta, and positron radiation is used for medical imaging, nondestructive testing, and a variety of industrial gauges. Radioactive tracers are used in medical and industrial applications, as well as biological and radiation chemistry. Alpha radiation is used in static eliminators and smoke detectors. The sterilizing effects of ionizing radiation are useful for cleaning medical instruments, food irradiation, and the sterile insect technique. Measurements of carbon-14, can be used to date the remains of long-dead organisms (such as wood that is thousands of years old).

Sources of Radiation

Ionizing radiation is generated through nuclear reactions, nuclear decay, by very high temperature, or via acceleration of charged particles in electromagnetic fields. Natural sources include the sun, lightning and supernova explosions. Artificial sources include nuclear reactors, particle accelerators, and x-ray tubes.

The United Nations Scientific Committee on the Effects of Atomic Radiation (UNSCEAR) itemized types of human exposures.

Type of radiation exposures		
Public exposure		
Natural Sources	Normal occurrences	Cosmic radiation
		Terrestrial radiation
	Enhanced sources	Metal mining and smelting
		Phosphate industry
		Coal mining and power production from coal
		Oil and gas drilling
		Rare earth and titanium dioxide industries
		Zirconium and ceramics industries
		Application of radium and thorium
		Other exposure situations
Man-made sources	Peaceful purposes	Nuclear power production
		Transport of nuclear and radioactive material
		Application other than nuclear power
	Military purposes	Nuclear tests
		Residues in the environment. Nuclear fallout
Historical situations		
Exposure from accidents		
Occupational radiation exposure		
Natural Sources		Cosmic ray exposures of aircrew and space crew
		Exposures in extractive and processing industries
		Gas and oil extraction industries
		Radon exposure in workplaces other than mines
Man-made sources	Peaceful purposes	Nuclear power industries
		Medical uses of radiation
		Industrial uses of radiation
		Miscellaneous uses
	Military purposes	Other exposed workers

The International Commission on Radiological Protection manages the International System of Radiological Protection, which sets recommended limits for dose uptake.

Background Radiation

Background radiation comes from both natural and man-made sources.

The global average exposure of humans to ionizing radiation is about 3 mSv (0.3 rem) per year, 80% of which comes from nature. The remaining 20% results from exposure to man-made radiation sources, primarily from medical imaging. Average man-made exposure is much higher in developed countries, mostly due to CT scans and nuclear medicine.

Natural background radiation comes from five primary sources: cosmic radiation, solar radiation, external terrestrial sources, radiation in the human body, and radon.

The background rate for natural radiation varies considerably with location, being as low as 1.5 mSv/a (1.5 mSv per year) in some areas and over 100 mSv/a in others. The highest level of purely natural radiation recorded on the Earth's surface is 90 μGy/h (0.8 Gy/a) on a Brazilian black beach composed of monazite. The highest background radiation in an inhabited area is found in Ramsar, primarily due to naturally radioactive limestone used as a building material. Some 2000 of the most exposed residents receive an average radiation dose of 10 mGy per year, (1 rad/yr) ten times more than the ICRP recommended limit for exposure to the public from artificial sources. Record levels were found in a house where the effective radiation dose due to external radiation was 135 mSv/a, (13.5 rem/yr) and the committed dose from radon was 640 mSv/a (64.0 rem/yr). This unique case is over 200 times higher than the world average background radiation. Despite the high levels of background radiation that the residents of Ramsar receive there is no compelling evidence that they experience a greater health risks. The ICRP recommendations are conservative limits and may represent an over representation of the actual health risk. Generally radiation safety organization recommend the most conservative limits assuming it is best to err on the side of caution. This level of caution is appropriate but should not be used to create fear about background radiation danger. Radiation danger from background radiation may be a serious threat but is more likely a small overall risk compared to all other factors in the environment.

Cosmic Radiation

The Earth, and all living things on it, are constantly bombarded by radiation from outside our solar system. This cosmic radiation consists of relativistic particles: positively charged nuclei (ions) from 1 amu protons (about 85% of it) to 26 amu iron nuclei and even beyond. (The high-atomic number particles are called HZE ions.) The energy of this radiation can far exceed that which humans can create, even in the largest particle accelerators. This radiation interacts in the atmosphere to create secondary radiation that rains down, including x-rays, muons, protons, antiprotons, alpha particles, pions, electrons, positrons, and neutrons.

The dose from cosmic radiation is largely from muons, neutrons, and electrons, with a dose rate that varies in different parts of the world and based largely on the geomagnetic field, altitude, and solar cycle. The cosmic-radiation dose rate on airplanes is so high that, according to the United Nations UNSCEAR 2000 Report, airline flight crew workers receive more dose on average than any other worker, including those in nuclear power plants. Airline crews receive more cosmic rays if they routinely work flight routes that take them close to the North or South pole at high altitudes, where this type of radiation is maximal.

Cosmic rays also include high-energy gamma rays, which are far beyond the energies produced by solar or human sources.

External Terrestrial Sources

Most materials on Earth contain some radioactive atoms, even if in small quantities. Most of the dose received from these sources is from gamma-ray emitters in building materials, or rocks and soil when outside. The major radionuclides of concern for terrestrial radiation are isotopes of potassium, uranium, and thorium. Each of these sources has been decreasing in activity since the formation of the Earth.

Internal Radiation Sources

All earthly materials that are the building-blocks of life contain a radioactive component. As humans, plants, and animals consume food, air, and water, an inventory of radioisotopes builds up within the organism. Some radionuclides, like potassium-40, emit a high-energy gamma ray that can be measured by sensitive electronic radiation measurement systems. These internal radiation sources contribute to an individual's total radiation dose from natural background radiation.

Radon

An important source of natural radiation is radon gas, which seeps continuously from bedrock but can, because of its high density, accumulate in poorly ventilated houses.

Radon-222 is a gas produced by the decay of radium-226. Both are a part of the natural uranium decay chain. Uranium is found in soil throughout the world in varying concentrations. Among non-smokers, radon is the largest cause of lung cancer and, overall, the second-leading cause.

Radiation Exposure

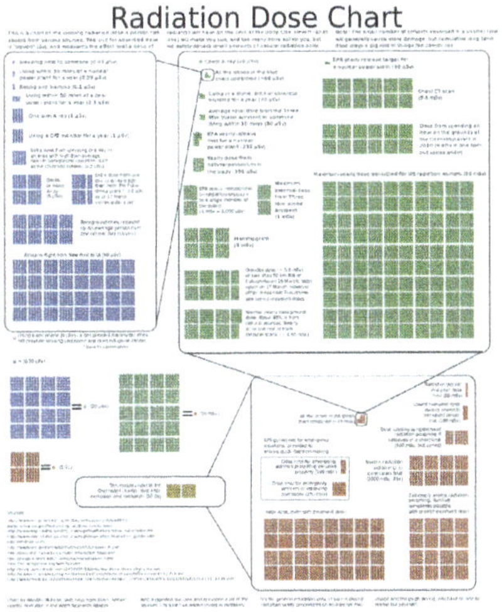

Various doses of radiation in sieverts, ranging from trivial to lethal.

There are three standard ways to limit exposure:

1. Time: For people exposed to radiation in addition to natural background radiation, limiting or minimizing the exposure time will reduce the dose from the radiation source.

2. Distance: Radiation intensity decreases sharply with distance, according to an inverse-square law (in an absolute vacuum).

3. Shielding: Air or skin can be sufficient to substantially attenuate alpha and beta radiation. Barriers of lead, concrete, or water are often used to give effective protection from more penetrating particles such as gamma rays and neutrons. Some radioactive materials are stored or handled underwater or by remote control in rooms constructed of thick concrete

or lined with lead. There are special plastic shields that stop beta particles, and air will stop most alpha particles. The effectiveness of a material in shielding radiation is determined by its half-value thicknesses, the thickness of material that reduces the radiation by half. This value is a function of the material itself and of the type and energy of ionizing radiation. Some generally accepted thicknesses of attenuating material are 5 mm of aluminum for most beta particles, and 3 inches of lead for gamma radiation.

These can all be applied to natural and man-made sources. For man-made sources the use of Containment is a major tool in reducing dose uptake and is effectively a combination of shielding and isolation from the open environment. Radioactive materials are confined in the smallest possible space and kept out of the environment such as in a hot cell (for radiation) or glove box (for contamination). Radioactive isotopes for medical use, for example, are dispensed in closed handling facilities, usually gloveboxes, while nuclear reactors operate within closed systems with multiple barriers that keep the radioactive materials contained. Work rooms, hot cells and gloveboxes have slightly reduced air pressures to prevent escape of airborne material to the open environment.

In nuclear conflicts or civil nuclear releases civil defense measures can help reduce exposure of populations by reducing ingestion of isotopes and occupational exposure . One is the issue of potassium iodide (KI) tablets, which blocks the uptake of radioactive iodine (one of the major radioisotope products of nuclear fission) into the human thyroid gland.

Occupational Exposure

Occupationally exposed individuals are controlled within the regulatory framework of the country they work in, and in accordance with any local nuclear licence constraints. These are usually based on the recommendations of the ICRP. The International Commission on Radiological Protection recommends limiting artificial irradiation. For occupational exposure, the limit is 50 mSv in a single year with a maximum of 100 mSv in a consecutive five-year period.

The radiation exposure of these individuals is carefully monitored with the use of dosimeters and other radiological protection instruments which will measure radioactive particulate concentrations, area gamma dose readings and radioactive contamination. A legal record of dose is kept.

Examples of activities where occupational exposure is a concern include:

- Airline crew (the most exposed population)
- Industrial radiography
- Medical radiology and nuclear medicine
- Uranium mining
- Nuclear power plant and nuclear fuel reprocessing plant workers
- Research laboratories (government, university and private)

Some human-made radiation sources affect the body through direct radiation, known as effective dose (radiation) while others take the form of radioactive contamination and irradiate the body from within. The latter is known as committed dose.

Public Exposure

Medical procedures, such as diagnostic X-rays, nuclear medicine, and radiation therapy are by far the most significant source of human-made radiation exposure to the general public. Some of the major radionuclides used are I-131, Tc-99m, Co-60, Ir-192, and Cs-137. The public also is exposed to radiation from consumer products, such as tobacco (polonium-210), combustible fuels (gas, coal, etc.), televisions, luminous watches and dials (tritium), airport X-ray systems, smoke detectors (americium), electron tubes, and gas lantern mantles (thorium).

Of lesser magnitude, members of the public are exposed to radiation from the nuclear fuel cycle, which includes the entire sequence from processing uranium to the disposal of the spent fuel. The effects of such exposure have not been reliably measured due to the extremely low doses involved. Opponents use a cancer per dose model to assert that such activities cause several hundred cases of cancer per year, an application of the widely accepted Linear no-threshold model (LNT).

The International Commission on Radiological Protection recommends limiting artificial irradiation to the public to an average of 1 mSv (0.001 Sv) of effective dose per year, not including medical and occupational exposures.

In a nuclear war, gamma rays from both the initial weapon explosion and fallout would be the sources of radiation exposure.

Spaceflight

Massive particles are a concern for astronauts outside the earth's magnetic field who would receive solar particles from solar proton events (SPE) and galactic cosmic rays from cosmic sources. These high-energy charged nuclei are blocked by Earth's magnetic field but pose a major health concern for astronauts traveling to the moon and to any distant location beyond the earth orbit. Highly charged HZE ions in particular are known to be extremely damaging, although protons make up the vast majority of galactic cosmic rays. Evidence indicates past SPE radiation levels that would have been lethal for unprotected astronauts.

Air Travel

Air travel exposes people on aircraft to increased radiation from space as compared to sea level, including cosmic rays and from solar flare events. Software programs such as Epcard, CARI, SIEVERT, PCAIRE are attempts to simulate exposure by aircrews and passengers. An example of a measured dose (not simulated dose) is 6 μSv per hour from London Heathrow to Tokyo Narita on a high-latitude polar route. However, dosages can vary, such as during periods of high solar activity. The United States FAA requires airlines to provide flight crew with information about cosmic radiation, and an International Commission on Radiological Protection recommendation for the general public is no more than 1 mSv per year. In addition, many airlines do not allow pregnant flightcrew members, to comply with a European Directive. The FAA has a recommended limit of 1 mSv total for a pregnancy, and no more than 0.5 mSv per month.

Radiation Hazard Warning Signs

Hazardous levels of ionizing radiation are signified by the trefoil sign on a yellow background. These are usually posted at the boundary of a radiation controlled area or in any place where radiation levels are significantly above background due to human intervention.

The red ionizing radiation warning symbol (ISO 21482) was launched in 2007, and is intended for IAEA Category 1, 2 and 3 sources defined as dangerous sources capable of death or serious injury, including food irradiators, teletherapy machines for cancer treatment and industrial radiography units. The symbol is to be placed on the device housing the source, as a warning not to dismantle the device or to get any closer. It will not be visible under normal use, only if someone attempts to disassemble the device. The symbol will not be located on building access doors, transportation packages or containers.

Ionizing radiation hazard symbol

2007 ISO radioactivity danger symbol intended for IAEA Category 1, 2 and 3 sources defined as dangerous sources capable of death or serious injury.

Alpha Radioactivity

Alpha radiation is another name for the alpha particles emitted in the type of radioactive decay called alpha decay

Alpha particles are composite particles consisting of two protons and two neutrons tightly bound together below figure. They are emitted from the nucleus of some radionuclides during a form of radioactive decay, called alpha-decay. An alpha-particle is identical to the nucleus of a normal (atomic mass 4) helium atom or a doubly ionised helium atom.

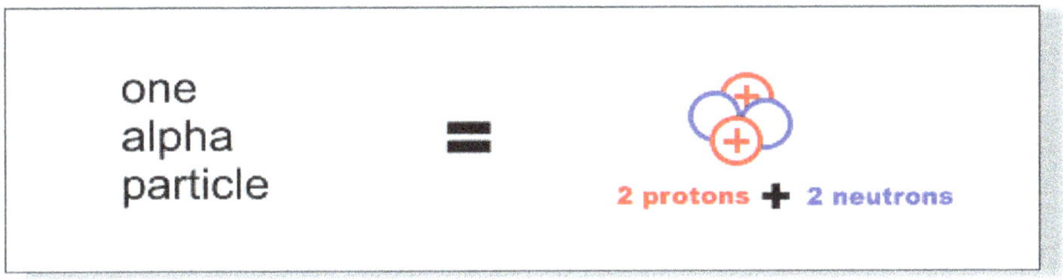

Alpha particles are relatively slow and heavy compared with other forms of nuclear radiation. The particles travels at 5 to 7 percent of the speed of light or 20,000,000 metres per second and has a mass approximately equivalent to 4 protons.

Ernest Rutherford, a New Zealand scientist, distinguished and named alpha rays in 1909. By measuring the charge and mass of alpha particles he discovered that they are the same as the nuclei of ordinary helium atoms. Rutherford's studies contributed to our understanding of the atom and its nucleus through the Rutherford-Bohr planetary model of the atom.

Causes of Some Radionuclides to Emit Alpha Particles

Alpha-decay occurs mainly in the radioactive decay of the heavier elements, particularly in those members of the natural decay series heavier than lead (atomic number 82), such as uranium and thorium. Alpha-particles are emitted with one of a few discrete energies characteristic of the radionuclide from which they were emitted. These energies can be used to identify the radionuclide involved.

Alpha-decay occurs when the ratio of neutrons to protons in the nucleus is low. For example: Polonium-210 has 126 neutrons and 84 protons, a ratio of 1.50 to 1. Following radioactive decay by the emission of an alpha particle, the ratio becomes 124 neutrons to 82 protons, or 1.51 to 1 below figure.

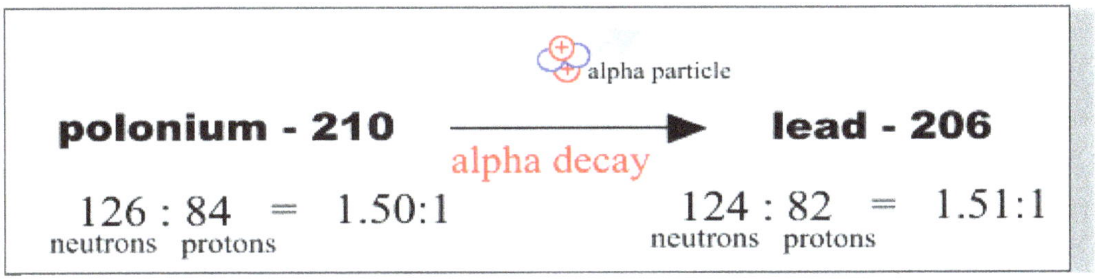

Behavior to Alpha Particles in the Environment

Alpha particles are highly ionising because of their double positive charge, large mass (compared to a beta particle)and because they are relatively slow. They can cause multiple ionisations within a very small distance. This gives them the potential to do much more biological damage for the same amount of deposited energy. Alpha-articles, because they are highly ionising, are unable to penetrate very far through matter and are brought to rest by a few centimetres of air or less than a tenth of a millimetre of living tissue in below figure.

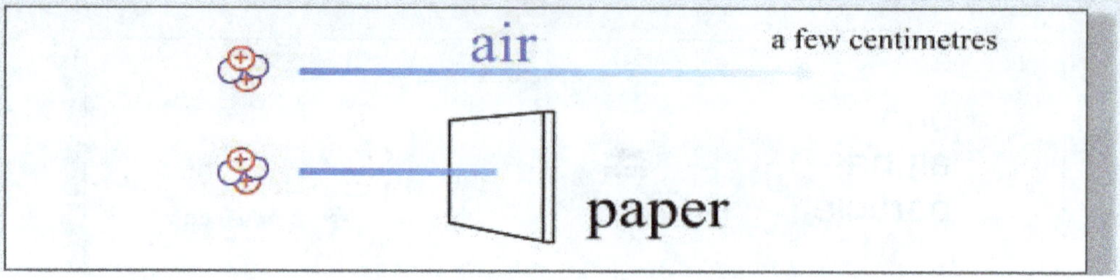

Alpha particles (or alpha-rays) were the first nuclear radiation to be specifically identified and hence their name, alpha. Beta and gamma-rays were identified soon after.

Effects of Exposure to Alpha Particles

Many alpha emitters occur naturally in the environment. For example, alpha particles are given off by uranium-238, radium-226, and other members of the uranium decay series which are present in varying amounts in nearly all rocks, soils, and water.

Alpha particles can't penetrate the normal layer of dead cells on the outside of our skin but could damage the cornea of the eye. Alpha-particle radiation is normally only a safety concern if the radioactive decay occurs in an atom that is already inside the body or inside a cell. Alpha-particle emitters are particularly dangerous if inhaled, ingested, or if they enter a wound.

Shielding of Alpha Radiation

The shielding of alpha radiation alone does not pose a difficult problem. On the other hand alpha radioactive nuclides can lead to serious health hazards when they are ingested or inhaled (internal contamination). When they are ingested or inhaled, the alpha particles from their decay significantly harm the internal living tissue. Moreover pure alpha radiation is very rare, alpha decay is frequently accompanied by gamma radiation which shielding is another issue.

Beta Radiation

Beta radiation consist of free electrons or positrons at relativistic speeds. These particles are known as the beta particles. Beta particles are high-energy, high-speed electrons or positrons emitted by certain fission fragments or by certain primordial radioactive nuclei such as potassium-40. The beta particles are a form of ionizing radiation also known as beta rays. The production of beta particles is termed beta decay. There are two forms of beta decay, the electron decay (β^- decay) and the positron decay (β^+ decay). In a nuclear reactor occurs especially the β^- decay, because the common feature of the fission products is an excess of neutrons An unstable fission fragment with the excess of neutrons undergoes β^- decay, where the neutron is converted into a proton, an electron, and an electron antineutrino.

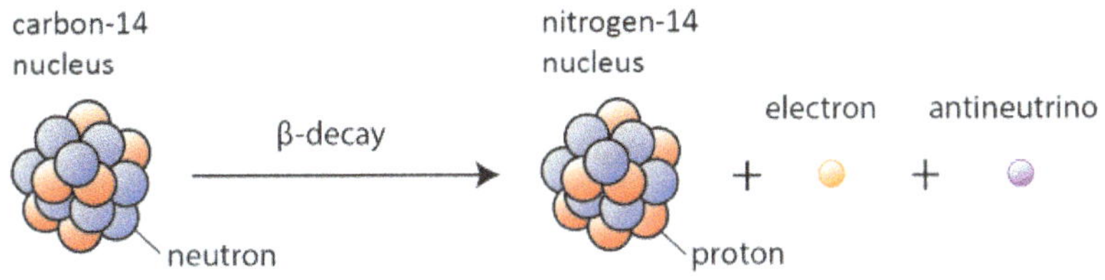

Characteristics of Beta Radiation

Key characteristics of beta radiation are summarized in following points:

- Beta particles are energetic electrons, they are relatively light and carry a single negative charge.
- Their mass is equal to the mass of the orbital electrons with which they are interacting and unlike the alpha particle a much larger fraction of its kinetic energy can be lost in a single interaction.

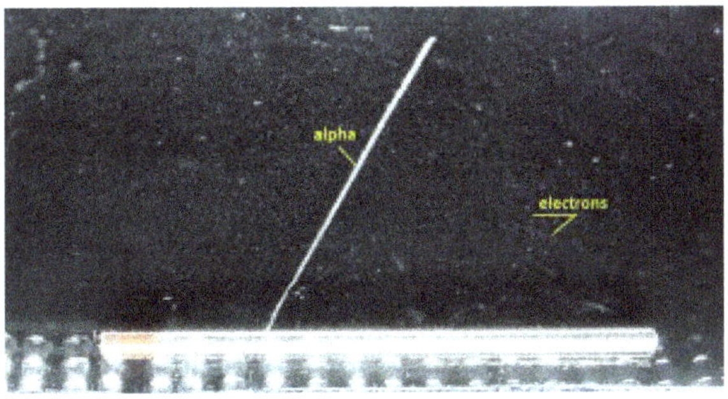

- Their path is not so straightforward. The beta particles follow a very zig-zag path through absorbing material. This resulting path of particle is longer than the linear penetration (range) into the material.
- Since they have very low mass, beta particles reach mostly relativistic energies.
- Beta particles also differ from other heavy charged particles in the fraction of energy lost by radiative process known as the bremsstrahlung. Therefore for high energy beta radiation shielding dense materials are inappropriate.
- When the beta particle moves faster than the speed of light (phase velocity) in the material it generates a shock wave of electromagnetic radiation known as the Cherenkov radiation.

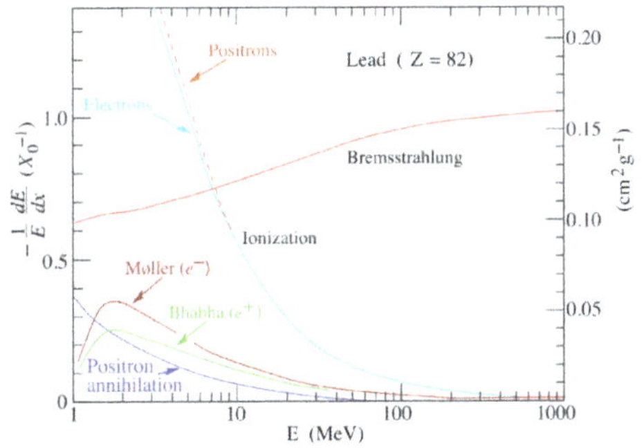

- The beta emission has the continuous spectrum.
- A 1 MeV beta particle can travel approximately 3.5 meters in air.
- Due to the presence of the bremsstrahlung low atomic number (Z) materials are appropriate as beta particle shields.

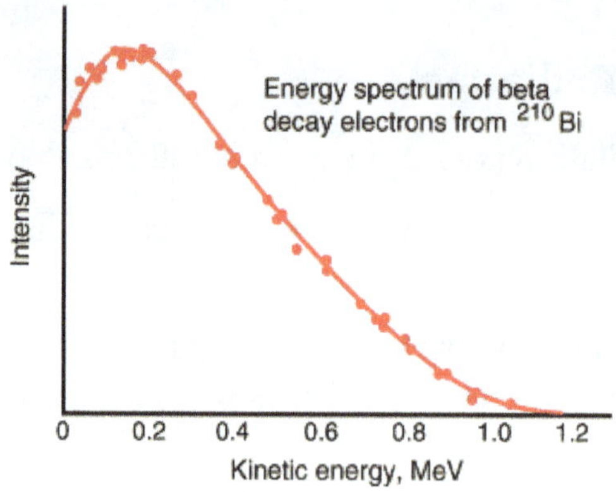

hielding of Beta Radiation – Electrons

Shielding of Beta Radiation – Electrons

Beta radiation ionizes matter weaker than alpha radiation. On the other hand the ranges of beta particles are longer and depends strongly on initial kinetic energy of particle. Some have enough energy to be of concern regarding external exposure. A 1 MeV beta particle can travel approximately 3.5 meters in air. Such beta particles can penetrate into the body and deposit dose to internal structures near the surface. Therefore greater shielding than in case of alpha radiation is required.

Materials with low atomic number Z are appropriate as beta particle shields. With high Z materials the bremsstrahlung (secondary radiation – X-rays) is associated. This radiation is created during slowing down of beta particles while they travel in a very dense medium. Heavy clothing, thick cardboard or thin aluminium plate will provide protection from beta radiation and prevents of production of the bremsstrahlung.

Shielding of Beta Radiation – Positrons

The coulomb forces that constitute the major mechanism of energy loss for electrons are present for either positive or negative charge on the particle and constitute the major mechanism of energy loss also for positrons. Whatever the interaction involves a repulsive or attractive force between the incident particle and orbital electron (or atomic nucleus), the impulse and energy transfer for particles of equal mass are about the same. Therefore positrons interact similarly with matter when they are energetic. The track of positrons in material is similar to the track of electrons. Even their specific energy loss and range are about the same for equal initial energies.

At the end of their path, positrons differ significantly from electrons. When a positron (antimatter particle) comes to rest, it interacts with an electron (matter particle), resulting in the annihilation of the both particles and the complete conversion of their rest mass to pure energy (according to the E=mc² formula) in the form of two oppositely directed 0.511 MeV gamma rays (photons).

Therefore any positron shield have to include also a gamma ray shield. In order to minimize the bremsstrahlung a multi-layered radiation shield is appropriate. Material for the first layer must fulfill the requirements for negative beta radiation shielding. First layer of such shield may be for example a thin aluminum plate (to shield positrons), while the second layer of such shield may be a dense material such as lead or depleted uranium.

β^- decay

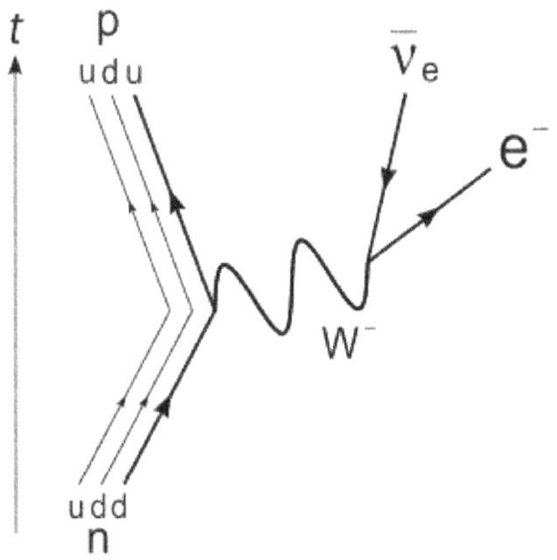

The leading-order Feynman diagram for β^- decay of a neutron into a proton, electron, and electron antineutrino via anintermediate W^-boson

In β^- decay, the weak interaction converts an atomic nucleus into a nucleus with atomic number increased by one, while emitting an electron (e^-) and an electron antineutrino (ν_e). β^- decay generally occurs in neutron-rich nuclei. The generic equation is:

$${}^{A}_{Z}X \rightarrow {}^{A}_{Z+1}X' + e^- + \bar{\nu}_e$$

where *A* and *Z* are the mass number and atomic number of the decaying nucleus, and X and X' are the initial and final elements, respectively.

Another example is when the free neutron (${}^{1}_{0}n$) decays by β^- decay into a proton (p):

$$n \rightarrow p + e^- + \bar{\nu}_e.$$

At the fundamental level (as depicted in the Feynman diagram on the right), this is caused by the conversion of the negatively charged ($-\frac{1}{3}$e) down quark to the positively charged ($+\frac{2}{3}$e) up quark by emission of a W^-boson; the W^-boson subsequently decays into an electron and an electron antineutrino:

$$d \rightarrow u + e^- + \bar{\nu}_e.$$

β^+ decay

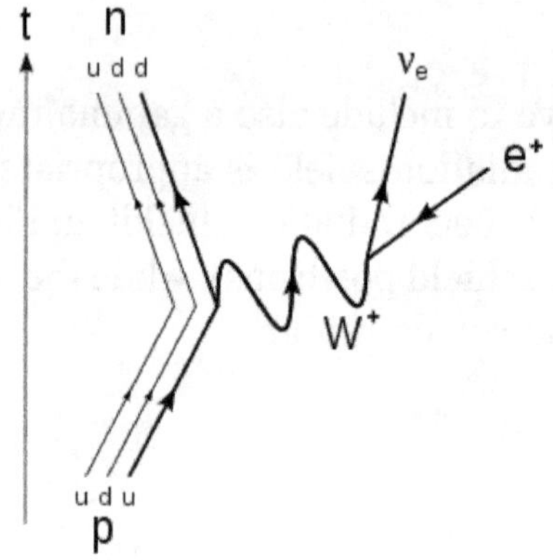

The leading-order Feynman diagram for β^+ decay of a proton into a neutron, positron, and electron neutrino via an intermediate W^+ boson.

In β^+ decay, or "positron emission", the weak interaction converts an atomic nucleus into a nucleus with atomic number decreased by one, while emitting a positron (e^+) and an electron neutrino (ν_e). β^+ decay generally occurs in proton-rich nuclei. The generic equation is:

$${}^{A}_{Z}X \rightarrow {}^{A}_{Z-1}X' + e^+ + \nu_e$$

This may be considered as the decay of a proton inside the nucleus to a neutron

$$p \rightarrow n + e^+ + \nu_e$$

However, β^+ decay cannot occur in an isolated proton because it requires energy due to the mass of the neutron being greater than the mass of the proton.

β^+decay can only happen inside nuclei when the daughter nucleus has a greater binding energy (and therefore a lower total energy) than the mother nucleus. The difference between these energies goes into the reaction of converting a proton into a neutron, a positron and a neutrino and into the kinetic energy of these particles. This process is opposite to negative beta decay, in that the weak interaction converts a proton into a neutron by converting an up quark into a down quark resulting in the emission of a W^+or the absorption of a W^-.

Electron Capture (K-capture)

In all cases where β^+ decay (positron emission) of a nucleus is allowed energetically, so too is electron capture allowed. This is a process during which a nucleus captures one of its atomic electrons, resulting in the emission of a neutrino:

$${}^{A}_{Z}X + e^- \rightarrow {}^{A}_{Z-1}X' + \nu_e$$

An example of electron capture is one of the decay modes of krypton-81 into bromine-81:

$${}^{81}_{36}Kr + e^- \rightarrow {}^{81}_{35}Br + \nu_e$$

All emitted neutrinos are of the same energy. In proton-rich nuclei where the energy difference between the initial and final states is less than $2m_ec^2$,

β^+ decay is not energetically possible, and electron capture is the sole decay mode.

If the captured electron comes from the innermost shell of the atom, the K-shell, which has the highest probability to interact with the nucleus, the process is called K-capture. If it comes from the L-shell, the process is called L-capture, etc.

Electron capture is a competing (simultaneous) decay process for all nuclei that can undergo β^+ decay. The converse, however, is not true: electron capture is the *only* type of decay that is allowed in proton-rich nuclides that do not have sufficient energy to emit a positron and neutrino.

Nuclear Transmutation

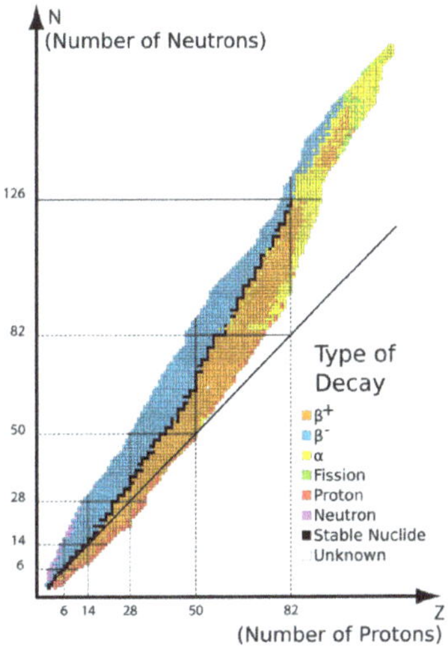

If the proton and neutron are part of an atomic nucleus, the above described decay processes transmute one chemical element into another. For example:

$^{137}_{55}Cs$			→	$^{137}_{56}Ba$	+	e^-	+	ν_e	(beta minus decay)
$^{22}_{11}Na$			→	$^{22}_{10}Ne$	+	e^+	+	ν_e	(beta plus decay)
$^{22}_{11}Na$	+	e^-	→	$^{22}_{10}Ne$	+	ν_e			(electron capture)

Beta decay does not change the number (*A*) of nucleons in the nucleus, but changes only its charge *Z*. Thus the set of all nuclides with the same *A* can be introduced; these *isobaric* nuclides may turn into each other via beta decay. For a given *A* there is one that is most stable. It is said to be beta stable, because it presents a local minima of the mass excess: if such a nucleus has (*A*, *Z*) numbers, the neighbour nuclei (*A*, *Z*−1) and (*A*, *Z*+1) have higher mass excess and can beta decay into (*A*, *Z*), but not vice versa. For all odd mass numbers *A*, there is only one known beta-stable isobar. For even *A*, there are up to three different beta-stable isobars experimentally known; for example, $^{96}_{40}Zr$, $^{96}_{42}Mo$, and $^{96}_{44}Ru$ are all beta-stable. There are about 355 known beta-decay stable nuclides.

Competition of Beta Decay Types

Usually unstable nuclides are clearly either "neutron rich" or "proton rich", with the former undergoing beta decay and the latter undergoing electron capture (or more rarely, due to the higher energy requirements, positron decay). However, in a few cases of odd-proton, odd-neutron radionuclides, it may be energetically favorable for the radionuclide to decay to an even-proton, even-neutron isobar either by undergoing beta-positive or beta-negative decay. An often-cited example is the single isotope ${}^{64}_{29}Cu$ (29 protons, 35 neutrons), which illustrates three types of beta decay in competition. Copper-64 has a half-life of about 12.7 hours. This isotope has one unpaired proton and one unpaired neutron, so either the proton or the neutron can decay. This particular nuclide (though not all nuclides in this situation) is almost equally likely to decay through proton decay by positron emission (18%) or electron capture (43%) to ${}^{64}_{28}Ni$, as it is through neutron decay by electron emission (39%) to ${}^{64}_{30}Zn$.

Stability of Naturally Occurring Nuclides

Most naturally occurring nuclides on earth are beta stable. Those that are not have half-lives ranging from under a second to periods of time significantly greater than the age of the universe. One common example of a long-lived isotope is the odd-proton odd-neutron nuclide ${}^{40}_{19}K$, which undergoes all three types of beta decay (β^-, $\beta^+$and electron capture) with a half-life of 1.277×10^9 years.

Conservation Rules for Beta Decay

Baryon Number is Conserved

$$B = \frac{n_q - n_{\bar{q}}}{3}$$

where

n_q is the number of constituent quarks, and

$n_{\bar{q}}$ is the number of constituent antiquarks.

Beta decay just changes neutron to proton or, in the case of positive beta decay (electron capture) proton to neutron so the number of individual quarks doesn't change. It is only the baryon flavor that changes, here labelled as the isospin.

Up and down quarks have total isospin $I = \frac{1}{2}$ and isospin projections

$$I_z = \begin{cases} \frac{1}{2} & \text{up quark} \\ -\frac{1}{2} & \text{down quark} \end{cases}$$

All other quarks have $I = 0$.

In general

$$I_z = \frac{1}{2}(n_u - n_d)$$

Lepton Number is Conserved

$$L \equiv n_\ell - n_{\bar{\ell}}$$

so all leptons have assigned a value of +1, antileptons −1, and non-leptonic particles 0.

$$\begin{array}{lcccccccc} & \text{n} & \rightarrow & \text{p} & + & \text{e}^- & + & \bar{\nu}_e \\ L: & 0 & = & 0 & + & 1 & - & 1 \end{array}$$

Angular Momentum

For allowed decays, the net orbital angular momentum is zero, hence only spin quantum numbers are considered.

The electron and antineutrino are fermions, spin-1/2 objects, therefore they may couple to total $S = 1$ (parallel) or $S = 0$ (anti-parallel).

For forbidden decays, orbital angular momentum must also be taken into consideration.

Energy Release

The Q value is defined as the total energy released in a given nuclear decay. In beta decay, Q is therefore also the sum of the kinetic energies of the emitted beta particle, neutrino, and recoiling nucleus. (Because of the large mass of the nucleus compared to that of the beta particle and neutrino, the kinetic energy of the recoiling nucleus can generally be neglected.) Beta particles can therefore be emitted with any kinetic energy ranging from 0 to Q. A typical Q is around 1 MeV, but can range from a few keV to a few tens of MeV.

Since the rest mass of the electron is 511 keV, the most energetic beta particles are ultrarelativistic, with speeds very close to the speed of light.

β⁻ decay

Consider the generic equation for beta decay

$${}^{A}_{Z}\text{X} \rightarrow {}^{A}_{Z+1}\text{X'} + \text{e}^- + \nu_e.$$

The Q value for this decay is

$$Q = \left[m_N\left({}^{A}_{Z}X\right) - m_N\left({}^{A}_{Z+1}X'\right)\right] - m_e - m_{\bar{\nu}_e}\Big]c^2.$$

where $m_N\left({}^{A}_{Z}X\right)$ is the mass of the nucleus of the ${}^{A}_{Z}\text{X}$ atom, m_e is the mass of the electron, and $m_{\bar{\nu}_e}$ is the mass of the electron antineutrino. In other words, the total energy released is the mass energy of the initial nucleus, minus the mass energy of the final nucleus, electron, and antineutrino. The mass of the nucleus m_N is related to the standard atomic mass m by

$$m\left({}_{Z}^{A}X\right)c^2 = m_N\left({}_{Z}^{A}X\right)c^2 + Zm_e c^2 - \sum_{i=1}^{Z} B_i.$$

That is, the total atomic mass is the mass of the nucleus, plus the mass of the electrons, minus the sum of all *electron* binding energies B_i for the atom. This equation is rearranged to find $m_N\left({}_{Z}^{A}X\right)$, and $m_N\left({}_{Z+1}^{A}X'\right)\big]$ is found similarly. Substituting these nuclear masses into the Q-value equation, while neglecting the nearly-zero antineutrino mass and the difference in electron binding energies, which is very small for high-Z atoms, we have:

$$Q = \left[m\left({}_{Z}^{A}X\right) - m\left({}_{Z-1}^{A}X'\right)\right]c^2$$

This energy is carried away as kinetic energy by the electron and neutrino.

Because the reaction will proceed only when the Q-value is positive, β^- decay can occur when the mass of atom ${}^{A}_{Z}X$is greater than the mass of atom ${}^{A}_{Z+1}X'$.

β^+ decay

The equations for β^+ decay are similar, with the generic equation:

$${}^{A}_{Z}X \rightarrow {}^{A}_{Z-1}X' + e^+ + \nu_e$$

giving

$$Q = \left[m_N\left({}_{Z}^{A}X\right) - m_N\left({}_{Z-1}^{A}X'\right) - m_e - m_{\nu_e}\right]c^2.$$

However, in this equation, the electron masses do not cancel, and we are left with:

$$Q = \left[m\left({}_{Z}^{A}X\right) - m\left({}_{Z-1}^{A}X'\right) - 2m_{\nu_e}\right]c^2$$

Because the reaction will proceed only when the Q-value is positive, β^+ decay can occur when the mass of atom ${}^{A}_{Z}X$exceeds that of ${}^{A}_{Z-1}X'$ by at least twice the mass of the electron.

Electron Capture

The analogous calculation for electron capture must take into account the binding energy of the electrons. This is because the atom will be left in an excited state after capturing the electron, and the binding energy of the captured innermost electron is significant. Using the generic equation for electron capture:

$${}^{A}_{Z}X + e^- \rightarrow {}^{A}_{Z-1}X' + \nu_e$$

we have

$$Q=\left[m_N\left({}^A_Z X\right)+m_e-m_N\left({}_{Z-1}^{\ A}X'\right)-m_{\nu_e}\right]c^2,$$

which simplifies to

$$Q=[m\left({}^A_Z X\right)-m\left({}_{Z-1}^{\ A}X'\right)]c^2-B_n,$$

where B_n is the binding energy of the captured electron.

Because the binding energy of the electron is much less than the mass of the electron, nuclei that can undergo β^+ decay can always also undergo electron capture, but the reverse is not true.

Beta Emission Spectrum

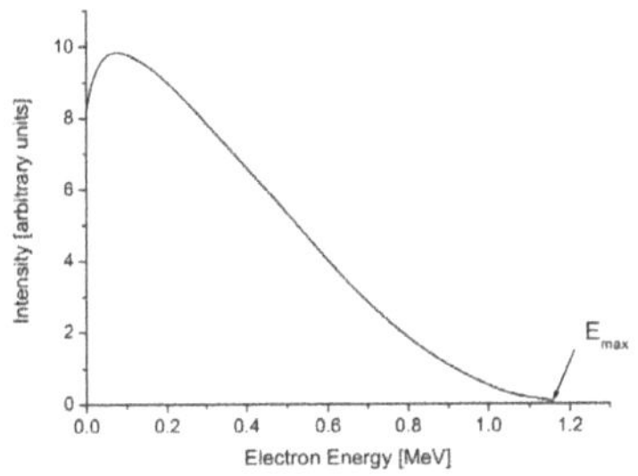

Beta spectrum of ^{210}Bi. E_{max}=Q=1.16 MeV is the maximum energy

Beta decay can be considered as a perturbation as described in quantum mechanics, and thus Fermi's Golden Rule can be applied. This leads to an expression for the kinetic energy spectrum $N(T)$ of emitted betas as follows:

$$N(T)=C_L(T)F(Z,T)pE(Q-T)^2$$

where T is the kinetic energy, C_L is a shape function that depends on the forbiddenness of the decay (it is constant for allowed decays), $F(Z, T)$ is the Fermi Function with Z the charge of the final-state nucleus, $E=T+mc^2$ is the total energy, $p=\sqrt{(E\ \ c)^2-(mc)^2}$ is the momentum, and Q is the Q value of the decay. The kinetic energy of the emitted neutrino is given approximately by Q minus the kinetic energy of the beta.

As an example, the beta decay spectrum of ^{210}Bi (originally called RaE) is shown to the right.

Fermi Function

The Fermi function that appears in the beta spectrum formula accounts for the Coulomb attraction / repulsion between the emitted beta and the final state nucleus. Approximating the associated wavefunctions to be spherically symmetric, the Fermi function can be analytically calculated to be:

$$F(Z,T)=\frac{2(1+S)}{\Gamma(1+2S)^2}(2p\rho)^{2S-2}e^{\pi\eta}\,|\Gamma(S+i\eta)|^2,$$

where $S = \sqrt{1 - \alpha^2 Z^2}$ (α is the fine-structure constant), $\eta = \pm \alpha ZE / pc$ (+ for electrons, – for positrons), $\rho = r_N/\hbar$ (r_N is the radius of the final state nucleus), and Γ is the Gamma function.

For non-relativistic betas ($Q \ll m_e c^2$), this expression can be approximated by:

$$F(Z,T) \approx \frac{2\pi\eta}{1 - e^{-2\pi\eta}}.$$

Other approximations can be found in the literature.

Kurie Plot

A Kurie plot (also known as a Fermi–Kurie plot) is a graph used in studying beta decay developed by Franz N. D. Kurie, in which the square root of the number of beta particles whose momenta (or energy) lie within a certain narrow range, divided by the Fermi function, is plotted against beta-particle energy. It is a straight line for allowed transitions and some forbidden transitions, in accord with the Fermi beta-decay theory. The energy-axis (x-axis) intercept of a Kurie plot corresponds to the maximum energy imparted to the electron/positron (the decay's *Q*-value). With a Kurie plot one can find the limit on the effective mass of a neutrino.

Helicity (Polarization) of Neutrinos, Electrons and Positrons Emitted in Beta Decay

After the discovery of parity non-conservation, it was found that, in beta decay, electrons are emitted mostly with negative helicity, i.e., they move, naively speaking, like left-handed screws driven into a material (they have negative longitudinal polarization). Conversely, positrons have mostly positive helicity, i.e., they move like right-handed screws. Neutrinos (emitted in positron decay) have positive helicity, while antineutrinos (emitted in electron decay) have negative helicity.

The higher the energy of the particles, the higher their polarization.

Types of Beta Decay Transitions

Beta decays can be classified according to the angular momentum (*L*-value) and total spin (*S*-value) of the emitted radiation. Since total angular momentum must be conserved, including orbital and spin angular momentum, beta decay occurs by a variety of quantum state transitions to various nuclear angular momentum or spin states, known as "Fermi" or "Gamow-Teller" transitions. When beta decay particles carry no angular momentum (L=0), the decay is referred to as "allowed", otherwise it is "forbidden".

Other decay modes, which are rare, are known as bound state decay and double beta decay.

Fermi Transitions

A Fermi transition is a beta decay in which the spins of the emitted electron (positron) and anti-neutrino (neutrino) couple to total spin $S = 0$, leading to an angular momentum change $\Delta J = 0$ between the initial and final states of the nucleus (assuming an allowed transition). In the non-relativistic limit, the nuclear part of the operator for a Fermi transition is given by

$$\mathcal{O}_F = G_V \sum_a \hat{\tau}_{a\pm}$$

with G_V the weak vector coupling constant, $\tau_\pm$ the isospin raising and lowering operators, and a running over all protons and neutrons in the nucleus.

Gamow-Teller Transitions

A Gamow-Teller transition is a beta decay in which the spins of the emitted electron (positron) and anti-neutrino (neutrino) couple to total spin $S = 1$, leading to an angular momentum change $\Delta J =$ ü ± between the initial and final states of the nucleus (assuming an allowed transition). In this case, the nuclear part of the operator is given by

$$\mathcal{O}_{GT} = G_A \sum_a \hat{\sigma}_a \hat{\tau}_{a\pm}$$

with G_A the weak axial-vector coupling constant, and σ the spin Pauli matrices, which can produce a spin-flip in the decaying nucleon.

Forbidden Transitions

When $L > 0$, the decay is referred to as "forbidden". Nuclear selection rules require high L-values to be accompanied by changes in nuclear spin (J) and parity (π). The selection rules for the Lth forbidden transitions are:

$$\Delta J = L-1, L, L+1; \Delta\pi = (-1)^L,$$

where $\Delta\pi=1$ or -1 corresponds to no parity change or parity change, respectively. The special case of a transition between isobaric analogue states, where the structure of the final state is very similar to the structure of the initial state, is referred to as "superallowed" for beta decay, and proceeds very quickly. The following table lists the ΔJ and $\Delta\pi$ values for the first few values of L:

Forbiddenness	ΔJ	$\Delta\pi$
Superallowed	0	no
Allowed	0, 1	no
First forbidden	0, 1, 2	yes
Second forbidden	1, 2, 3	no
Third forbidden	2, 3, 4	yes

Rare Decay Modes

Bound-state β^- decay

A very small minority of free neutron decays (about four per million) are so-called "two-body decays", in which the proton, electron and antineutrino are produced, but the electron fails to gain the 13.6 eV energy necessary to escape the proton, and therefore simply remains bound to it, as a neutral hydrogen atom. In this type of beta decay, in essence all of the neutron decay energy is carried off by the antineutrino.

For fully ionized atoms (bare nuclei), it is possible in likewise manner for electrons to fail to escape the atom, and to be emitted from the nucleus into low-lying atomic bound states (orbitals). This cannot occur for neutral atoms with low-lying bound states which are already filled by electrons.

Bound-state β decays were predicted by Daudel, Jean, and Lecoin in 1947, and the phenomenon in fully ionized atoms was first observed for $^{163}Dy^{66+}$ in 1992 by Jung et al. of the Darmstadt Heavy-Ion Research group. Although neutral ^{163}Dy is a stable isotope, the fully ionized $^{163}Dy^{66+}$ undergoes β decay into the K and L shells with a half-life of 47 days.

Another possibility is that a fully ionized atom undergoes greatly accelerated β decay, as observed for ^{187}Re by Bosch et al., also at Darmstadt. Neutral ^{187}Re does undergo β decay with a half-life of 42×10^9 years, but for fully ionized $^{187}Re^{75+}$ this is shortened by a factor of 10^9 to only 32.9 years. For comparison the variation of decay rates of other nuclear processes due to chemical environment is less than 1%.

Double Beta Decay

Some nuclei can undergo double beta decay (ββ decay) where the charge of the nucleus changes by two units. Double beta decay is difficult to study, as the process has an extremely long half-life. In nuclei for which both β decay and ββ decay are possible, the rarer ββ decay process is effectively impossible to observe. However, in nuclei where β decay is forbidden but ββ decay is allowed, the process can be seen and a half-life measured. Thus, ββ decay is usually studied only for beta stable nuclei. Like single beta decay, double beta decay does not change A; thus, at least one of the nuclides with some given A has to be stable with regard to both single and double beta decay.

"Ordinary" double beta decay results in the emission of two electrons and two antineutrinos. If neutrinos are Majorana particles (i.e., they are their own antiparticles), then a decay known as neutrinoless double beta decay will occur. Most neutrino physicists believe that neutrinoless double beta decay has never been observed.

Gamma Radiation

Gamma ray or Gamma radiation are the electromagnetic radiation of the shortest wavelength and highest energy.

Gamma rays are produced in the disintegration of radioactive atomic nuclei and in the decay of certain subatomic particles. The commonly accepted definitions of the gamma-ray and X-ray regions of the electromagnetic spectrum include some wavelength overlap, with gamma-ray radiation having wavelengths that are generally shorter than a few tenths of an angstrom (10^{-10} metre) and gamma-ray photons having energies that are greater than tens of thousands of electron volts (eV). There is no theoretical upper limit to the energies of gamma-ray photons and no lower limit to gamma-ray wavelengths; observed energies presently extend up to a few trillion electron volts—these extremely high-energy photons are produced in astronomical sources through currently unidentified mechanisms.

The term *gamma ray* was coined by British physicist Ernest Rutherford in 1903 following early studies of the emissions of radioactive nuclei. Just as atoms have discrete energy levels associated with different configurations of the orbiting electrons, atomic nuclei have energy level structures determined by the configurations of the protons and neutrons that constitute the nuclei. While energy differences between atomic energy levels are typically in the 1- to 10-eV range, energy differences in nuclei usually fall in the 1-keV (thousand electron volts) to 10-MeV (million electron volts) range. When a nucleus makes a transition from a high-energy level to a lower-energy level, a photonis emitted to carry off the excess energy; nuclear energy-level differences correspond to photon wavelengths in the gamma-ray region.

When an unstable atomic nucleus decays into a more stable nucleus, the "daughter" nucleus is sometimes produced in an excited state. The subsequent relaxation of the daughter nucleus to a lower-energy state results in the emission of a gamma-ray photon. Gamma-ray spectroscopy, involving the precise measurement of gamma-ray photon energies emitted by different nuclei, can establish nuclear energy-level structures and allows for the identification of trace radioactive elements through their gamma-ray emissions. Gamma rays are also produced in the important process of pair annihilation, in which an electron and its antiparticle, a positron, vanish and two photons are created. The photons are emitted in opposite directions and must each carry 511 keV of energy—the rest mass energy of the electron and positron. Gamma rays can also be generated in the decay of some unstable subatomic particles, such as the neutral pion.

Gamma-ray photons, like their X-ray counterparts, are a form of ionizing radiation; when they pass through matter, they usually deposit their energy by liberating electrons from atoms and molecules. At the lower energy ranges, a gamma-ray photon is often completely absorbed by an atom and the gamma ray's energy transferred to a single ejected electron . Higher-energy gamma rays are more likely to scatter from the atomic electrons, depositing a fraction of their energy in each scattering event. Standard methods for the detection of gamma rays are based on the effects of the liberated atomic electrons in gases, crystals, and semiconductors.

Electrons and positrons produced simultaneously from individual gamma rays curl in opposite directions in the magnetic field of a bubble chamber. In the top example, the gamma ray has lost some energy to an atomic electron, which leaves the long track, curling left. The gamma rays do not leave tracks in the chamber, as they have no electric charge.

Gamma rays can also interact with atomic nuclei. In the process of pair production, a gamma-ray photon with an energy exceeding twice the rest mass energy of the electron (greater than 1.02 MeV), when passing close to a nucleus, is directly converted into an electron-positron pair. At even higher energies (greater than 10 MeV), a gamma ray can be directly absorbed by a nucleus, causing the ejection of nuclear particles or the splitting of the nucleus in a process known as photofission.

Medical applications of gamma rays include the valuable imaging technique of positron emission tomography (PET) and effective radiation therapies to treat cancerous tumours. In a PET scan, a short-lived positron-emitting radioactive pharmaceutical, chosen because of its participation in a particular physiological process (e.g., brain function), is injected into the body. Emitted positrons quickly combine with nearby electrons and, through pair annihilation, give rise to two 511-keV gamma rays traveling in opposite directions. After detection of the gamma rays, a computer-generated reconstruction of the locations of the gamma-ray emissions produces an image that highlights the location of the biological process being examined.

As a deeply penetrating ionizing radiation, gamma rays cause significant biochemical changes in living cells. Radiation therapies make use of this property to selectively destroy cancerous cells in small localized tumours. Radioactive isotopes are injected or implanted near the tumour; gamma rays that are continuously emitted by the radioactive nuclei bombard the affected area and arrest the development of the malignant cells.

Airborne surveys of gamma-ray emissions from the Earth's surface search for minerals containing trace radioactive elements such as uranium and thorium. Aerial and ground-based gamma-ray spectroscopy is employed to support geologic mapping, mineral exploration, and identification of environmental contamination. Gamma rays were first detected from astronomical sources in the 1960s, and gamma-ray astronomy is now a well-established field of research. As with the study of astronomical X-rays, gamma-ray observations must be made above the strongly absorbing atmosphere of the Earth—typically with orbiting satellites or high-altitude balloons. There are many intriguing and poorly understood astronomical gamma-ray sources, including powerful point sources tentatively identified as pulsars, quasars, and supernovaremnants. Among the most fascinating unexplained astronomical phenomena are so-called gamma-ray bursts—brief, extremely intense emissions from sources that are apparently isotropically distributed in the sky.

Sources

This animation tracks several gamma rays through space and time, from their emission in the jet of a distant blazar to their arrival in Fermi's Large Area Telescope (LAT).

Natural sources of gamma rays on Earth include gamma decay from naturally occurring radioisotopes such as potassium-40, and also as a secondary radiation from various atmospheric interactions with cosmic ray particles. Some rare terrestrial natural sources that produce gamma rays that are not of a nuclear origin, are lightning strikes and terrestrial gamma-ray flashes, which produce high energy emissions from natural high-energy voltages. Gamma rays are produced by a number of astronomical processes in which very high-energy electrons are produced. Such electrons produce secondary gamma rays by the mechanisms of bremsstrahlung, inverse Compton scattering and synchrotron radiation. A large fraction of such astronomical gamma rays are screened by Earth's atmosphere and must be detected by spacecraft. Notable artificial sources of gamma rays include fission, such as occurs in nuclear reactors, as well as high energy physics experiments, such as neutral pion decay and nuclear fusion.

A sample of gamma ray-emitting material that is used for irradiating or imaging is known as a gamma source. It is also called a radioactive source, isotope source, or radiation source, though these more general terms also apply to alpha- and beta-emitting devices. Gamma sources are usually sealed to prevent radioactive contamination, and transported in heavy shielding.

Radioactive Decay (Gamma Decay)

Gamma rays are produced during gamma decay, which normally occurs after other forms of decay occur, such as alpha or beta decay. An excited nucleus can decay by the emission of an α or β particle. The daughter nucleus that results is usually left in an excited state. It can then decay to a lower energy state by emitting a gamma ray photon, in a process called gamma decay.

The emission of a gamma ray from an excited nucleus typically requires only 10^{-12} seconds. Gamma decay may also follow nuclear reactions such as neutron capture, nuclear fission, or nuclear fusion. Gamma decay is also a mode of relaxation of many excited states of atomic nuclei following other types of radioactive decay, such as beta decay, so long as these states possess the necessary component of nuclear spin. When high-energy gamma rays, electrons, or protons bombard materials, the excited atoms emit characteristic "secondary" gamma rays, which are products of the creation of excited nuclear states in the bombarded atoms. Such transitions, a form of nuclear gamma fluorescence, form a topic in nuclear physics called gamma spectroscopy. Formation of fluorescent gamma rays are a rapid subtype of radioactive gamma decay.

In certain cases, the excited nuclear state that follows the emission of a beta particle or other type of excitation, may be more stable than average, and is termed a metastable excited state, if its decay takes (at least) 100 to 1000 times longer than the average 10^{-12} seconds. Such relatively long-lived excited nuclei are termed nuclear isomers, and their decays are termed isomeric transitions. Such nuclei have half-lifes that are more easily measurable, and rare nuclear isomers are able to stay in their excited state for minutes, hours, days, or occasionally far longer, before emitting a gamma ray. The process of isomeric transition is therefore similar to any gamma emission, but differs in that it involves the intermediate metastable excited state(s) of the nuclei. Metastable states are often characterized by high nuclear spin, requiring a change in spin of several units or more with gamma decay, instead of a single unit transition that occurs in only 10^{-12} seconds. The rate of gamma decay is also slowed when the energy of excitation of the nucleus is small.

An emitted gamma ray from any type of excited state may transfer its energy directly to any electrons, but most probably to one of the K shell electrons of the atom, causing it to be ejected from that atom, in a process generally termed the photoelectric effect (external gamma rays and ultraviolet rays may also cause this effect). The photoelectric effect should not be confused with the internal conversion process, in which a gamma ray photon is not produced as an intermediate particle (rather, a "virtual gamma ray" may be thought to mediate the process).

Decay Schemes

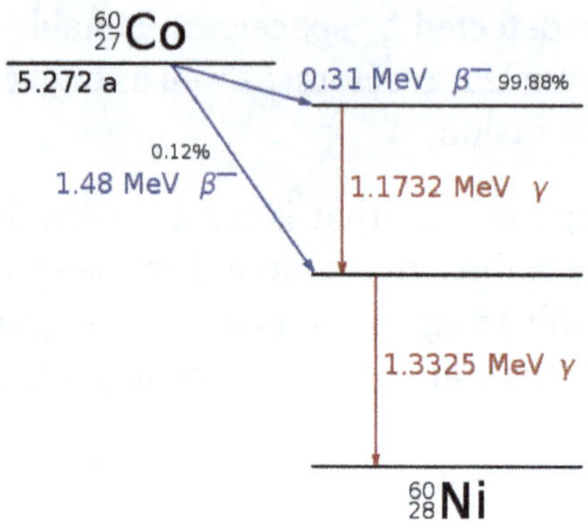

Radioactive decay scheme of ^{60} Co

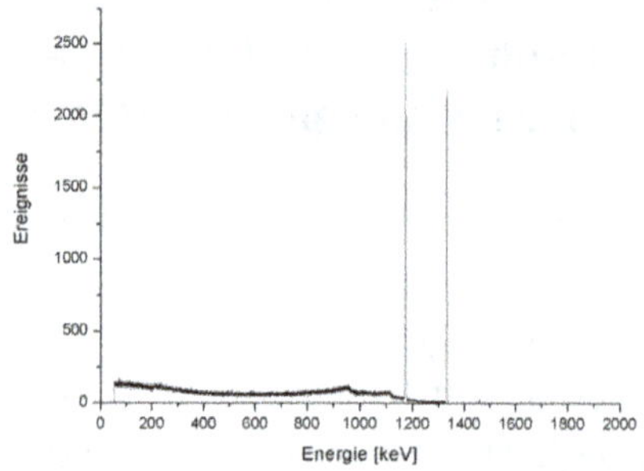

Gamma emission spectrum of cobalt-60

One example of gamma ray production due to radionuclide decay is the decay scheme for Cobalt 60, as illustrated in the accompanying diagram. First, ^{60} Co decays to excited ^{60} Ni by beta decay emission of an electron of 0.31 MeV. Then the excited ^{60} Ni decays to the ground state by emitting gamma rays in succession of 1.17 MeV followed by 1.33 MeV. This path is followed 99.88% of the time:

$^{60}_{27}Co$	→	$^{60}_{28}Ni^*$	+	e^-	+	ν_e	+	γ	+	1.17 MeV
$^{60}_{28}Ni^*$	→	$^{60}_{28}Ni$					+	γ	+	1.33 MeV

Another example is the alpha decay of ^{241} Am to form ^{237} Np ; which is followed by gamma emission. In some cases, the gamma emission spectrum of the daughter nucleus is quite simple, (e.g. ^{60} Co /^{60} Ni) while in other cases, such as with (^{241} Am /^{237} Np and ^{192} Ir /^{192} Pt), the gamma emission spectrum is complex, revealing that a series of nuclear energy levels exist.

Particle Physics

Gamma rays are produced in many processes of particle physics. Typically, gamma rays are the products of neutral systems which decay through electromagnetic interactions (rather than a weak or strong interaction). For example, in an electron–positron annihilation, the usual products are two gamma ray photons. If the annihilating electron and positron are at rest, each of the resulting gamma rays has an energy of ~ 511 keV and frequency of ~ 1.24×10^{20} Hz. Similarly, a neutral pion most often decays into two photons. Many other hadrons and massive bosons also decay electromagnetically. High energy physics experiments, such as the Large Hadron Collider, accordingly employ substantial radiation shielding. Because subatomic particles mostly have far shorter wavelengths than atomic nuclei, particle physics gamma rays are generally several orders of magnitude more energetic than nuclear decay gamma rays. Since gamma rays are at the top of the electromagnetic spectrum in terms of energy, all extremely high-energy photons are gamma rays; for example, a photon having the Planck energy would be a gamma ray.

Gamma Rays from Sources other than Radioactive Decay

A few gamma rays in astronomy are known to arise from gamma decay , but most do not.

Photons from astrophysical sources that carry energy in the gamma radiation range are often explicitly called gamma-radiation. In addition to nuclear emissions, they are often produced by sub-atomic particle and particle-photon interactions. Those include electron-positron annihilation, neutral pion decay, bremsstrahlung, inverse Compton scattering, and synchrotron radiation.

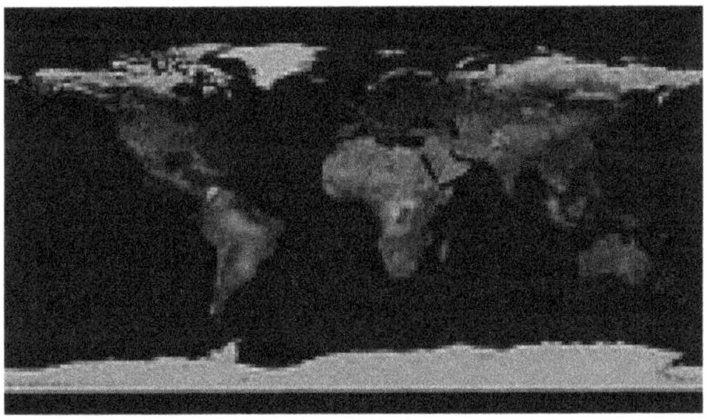

The red dots show some of the ~500 terrestrial gamma-ray flashes daily detected by the Fermi Gamma-ray Space Telescope through 2010.

Laboratory Sources

In October 2017, scientists from various European universities proposed a means for sources of GeV photons using lasers as exciters through a controlled interplay between the cascade and anomalous radiative trapping.

Terrestrial Thunderstorms

Thunderstorms can produce a brief pulse of gamma radiation called a terrestrial gamma-ray flash. These gamma rays are thought to be produced by high intensity static electric fields accelerating electrons, which then produce gamma rays by bremsstrahlung as they collide with and are slowed by atoms in the atmosphere. Gamma rays up to 100 MeV can be emitted by terrestrial thunderstorms, and were discovered by space-borne observatories. This raises the possibility of health risks to passengers and crew on aircraft flying in or near thunderclouds.

Cosmic Rays

Extraterrestrial, high energy gamma rays include the gamma ray background produced when cosmic rays (either high speed electrons or protons) collide with ordinary matter, producing pair-production gamma rays at 511 keV. Alternatively, bremsstrahlung are produced at energies of tens of MeV or more when cosmic ray electrons interact with nuclei of sufficiently high atomic number.

Image of entire sky in 100 MeV or greater gamma rays as seen by the EGRET instrument aboard the CGRO spacecraft. Bright spots within the galactic plane are pulsars while those above and below the plane are thought to be quasars.

Pulsars and Magnetars

The gamma ray sky is dominated by the more common and longer-term production of gamma rays that emanate from pulsars within the Milky Way. Sources from the rest of the sky are mostly quasars. Pulsars are thought to be neutron stars with magnetic fields that produce focused beams of radiation, and are far less energetic, more common, and much nearer sources (typically seen only in our own galaxy) than are quasars or the rarer gamma-ray burst sources of gamma rays. Pulsars have relatively long-lived magnetic fields that produce focused beams of relativistic speed charged particles, which emit gamma rays (bremsstrahlung) when those strike gas or dust in their nearby medium, and are decelerated. This is a similar mechanism to the production of high-energy photons in megavoltage radiation therapy machines. Inverse Compton scattering, in which charged particles (usually electrons) impart energy to low-energy photons boosting them to higher energy photons. Such impacts of photons on relativistic charged particle beams is another possible mechanism of gamma ray production. Neutron stars with a very high magnetic field (magnetars), thought to produce astronomical soft gamma repeaters, are another relatively long-lived star-powered source of gamma radiation.

Quasars and Active Galaxies

More powerful gamma rays from very distant quasars and closer active galaxies are thought to have a gamma ray production source similar to a particle accelerator. High energy electrons produced by the quasar, and subjected to inverse Compton scattering, synchrotron radiation, or bremsstrahlung, are the likely source of the gamma rays from those objects. It is thought that a supermassive black hole at the center of such galaxies provides the power source that intermittently destroys stars and focuses the resulting charged particles into beams that emerge from their rotational poles. When those beams interact with gas, dust, and lower energy photons they produce X-rays and gamma rays. These sources are known to fluctuate with durations of a few weeks, suggesting their relatively small size (less than a few light-weeks across). Such sources of gamma and X-rays are the most commonly visible high intensity sources outside our galaxy. They shine not in bursts, but relatively continuously when viewed with gamma ray telescopes. The power of a typical quasar is about 10^{40} watts, a small fraction of which is gamma radiation. Much of the rest is emitted as electromagnetic waves of all frequencies, including radio waves.

A hypernova. Artist's illustration showing the life of a massive star as nuclear fusion converts lighter elements into heavier ones. When fusion no longer generates enough pressure to counteract gravity, the star rapidly collapses to form a black hole. Theoretically, energy may be released during the collapse along the axis of rotation to form a long duration gamma-ray burst.

Gamma-ray Bursts

The most intense sources of gamma rays, are also the most intense sources of any type of electromagnetic radiation presently known. They are the "long duration burst" sources of gamma rays in astronomy ("long" in this context, meaning a few tens of seconds), and they are rare compared with the sources discussed above. By contrast, "short" gamma-ray bursts of two seconds or less, which are not associated with supernovae, are thought to produce gamma rays during the collision of pairs of neutron stars, or a neutron star and a black hole.

The so-called *long-duration* gamma-ray bursts produce a total energy output of about 10^{44} joules (as much energy as our Sun will produce in its entire life-time) but in a period of only 20 to 40 seconds. Gamma rays are approximately 50% of the total energy output. The leading hypotheses for the mechanism of production of these highest-known intensity beams of radiation, are inverse Compton scattering and synchrotron radiation from high-energy charged particles. These processes occur as relativistic charged particles leave the region of the event horizon of a newly formed black hole created during supernova explosion. The beam of particles moving at relativistic speeds are focused for a few tens of seconds by the magnetic field of the exploding hypernova. The fusion explosion of the hypernova drives the energetics of the process. If the narrowly directed beam happens to be pointed toward the Earth, it shines at gamma ray frequencies with such intensity, that it can be detected even at distances of up to 10 billion light years, which is close to the edge of the visible universe.

Properties

Penetration of Matter

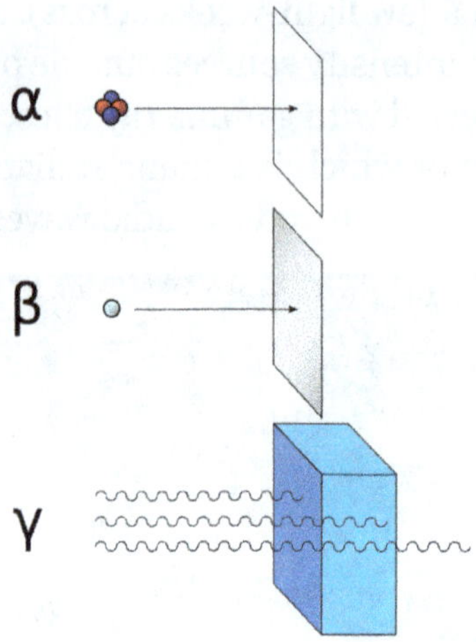

Alpha radiation consists of helium nuclei and is readily stopped by a sheet of paper. Beta radiation, consisting of electrons or positrons, is stopped by an aluminum plate, but gamma radiation requires shielding by dense material such as lead or concrete.

Due to their penetrating nature, gamma rays require large amounts of shielding mass to reduce them to levels which are not harmful to living cells, in contrast to alpha particles, which can be stopped by paper or skin, and beta particles, which can be shielded by thin aluminium. Gamma rays are best absorbed by materials with high atomic numbers and high density, which contribute to the total stopping power. Because of this, a lead (high Z) shield is 20–30% better as a gamma shield than an equal mass of another low-Z shielding material, such as aluminium, concrete, water, or soil; lead's major advantage is not in lower weight, but rather its compactness due to its higher density. Protective clothing, goggles and respirators can protect from internal contact with or ingestion of alpha or beta emitting particles, but provide no protection from gamma radiation from external sources.

The higher the energy of the gamma rays, the thicker the shielding made from the same shielding material is required. Materials for shielding gamma rays are typically measured by the thickness required to reduce the intensity of the gamma rays by one half (the half value layer or HVL). For example, gamma rays that require 1 cm (0.4″) of lead to reduce their intensity by 50% will also have their intensity reduced in half by 4.1 cm of granite rock, 6 cm (2½″) of concrete, or 9 cm (3½″) of packed soil. However, the mass of this much concrete or soil is only 20–30% greater than that of lead with the same absorption capability. Depleted uranium is used for shielding in portable gamma ray sources, but here the savings in weight over lead are larger, as portable sources' shape resembles a sphere to some extent, and the volume of a sphere is dependent on the cube of the radius; so a source with its radius cut in half will have its volume reduced by a factor of eight, which will more than compensate uranium's greater density (as well as reducing bulk). In a nuclear power plant, shielding can be provided by steel and concrete in the pressure and particle containment vessel, while water provides a radiation shielding of fuel rods during storage or transport into the reactor core. The loss of water or removal of a "hot" fuel assembly into the air would result in much higher radiation levels than when kept under water.

Light Interaction

High-energy (from 80 GeV to ~10 TeV) gamma rays arriving from far-distant quasars are used to estimate the extragalactic background light in the universe: The highest-energy rays interact more readily with the background light photons and thus the density of the background light may be estimated by analyzing the incoming gamma ray spectra.

Gamma Spectroscopy

Gamma spectroscopy is the study of the energetic transitions in atomic nuclei, which are generally associated with the absorption or emission of gamma rays. As in optical spectroscopy the absorption of gamma rays by a nucleus is especially likely (i.e., peaks in a "resonance") when the energy of the gamma ray is the same as that of an energy transition in the nucleus. In the case of gamma rays, such a resonance is seen in the technique of Mössbauer spectroscopy. In the Mössbauer effect the narrow resonance absorption for nuclear gamma absorption can be successfully attained by physically immobilizing atomic nuclei in a crystal. The immobilization of nuclei at both ends of a gamma resonance interaction is required so that no gamma energy is lost to the kinetic energy of recoiling nuclei at either the emitting or absorbing end of a gamma transition. Such loss of energy causes gamma ray resonance absorption to fail. However, when emitted gamma rays carry essentially all of the energy of the atomic nuclear de-excitation that produces them, this energy is also sufficient to excite the same energy state in a second immobilized nucleus of the same type.

Health Effects

Gamma rays cause damage at a cellular level and are penetrating, causing diffuse damage throughout the body. However, they are less ionising than alpha or beta particles, which are less penetrating.

Low levels of gamma rays cause a stochastic health risk, which for radiation dose assessment is defined as the *probability* of cancer induction and genetic damage. High doses produce deterministic effects, which is the *severity* of acute tissue damage that is certain to happen. These effects are compared to the physical quantity absorbed dose measured by the unit gray (Gy).

Body Response

When gamma radiation breaks DNA molecules, a cell may be able to repair the damaged genetic material, within limits. However, a study of Rothkamm and Lobrich has shown that this repair process works well after high-dose exposure but is much slower than in the case of a low-dose exposure.

Risk Assessment

The natural outdoor exposure in Great Britain ranges from 0.1 to 0.5 μSv/h with significant increase around known nuclear and contaminated sites. Natural exposure to gamma rays is about 1 to 2 mSv per year, and the average total amount of radiation received in one year per inhabitant in the USA is 3.6 mSv. There is a small increase in the dose, due to naturally occurring gamma radiation, around small particles of high atomic number materials in the human body caused by the photoelectric effect.

By comparison, the radiation dose from chest radiography (about 0.06 mSv) is a fraction of the annual naturally occurring background radiation dose. A chest CT delivers 5 to 8 mSv. A whole-body PET/CT scan can deliver 14 to 32 mSv depending on the protocol. The dose from fluoroscopy of the stomach is much higher, approximately 50 mSv (14 times the annual background).

An acute full-body equivalent single exposure dose of 1 Sv (1000 mSv) causes slight blood changes, but 2.0–3.5 Sv (2.0–3.5 Gy) causes very severe syndrome of nausea, hair loss, and hemorrhaging, and will cause death in a sizable number of cases—-about 10% to 35% without medical treatment. A dose of 5 Sv (5 Gy) is considered approximately the LD_{50} (lethal dose for 50% of exposed population) for an acute exposure to radiation even with standard medical treatment. A dose higher than 5 Sv (5 Gy) brings an increasing chance of death above 50%. Above 7.5–10 Sv (7.5–10 Gy) to the entire body, even extraordinary treatment, such as bone-marrow transplants, will not prevent the death of the individual exposed.

For low-dose exposure, for example among nuclear workers, who receive an average yearly radiation dose of 19 mSv, the risk of dying from cancer (excluding leukemia) increases by 2 percent. For a dose of 100 mSv, the risk increase is 10 percent. By comparison, risk of dying from cancer was increased by 32 percent for the survivors of the atomic bombing of Hiroshima and Nagasaki.

Units of Measurement and Exposure

The following table shows radiation quantities in SI and non-SI units:

Radiation related quantities					
Quantity	Unit	Symbol	Derivation	Year	SI equivalence
Activity (*A*)	curie	Ci	$3.7 \times 10^{10}\ s^{-1}$	1953	3.7×10^{10} Bq
	becquerel	Bq	s^{-1}	1974	SI
	rutherford	Rd	$10^{6}\ s^{-1}$	1946	1,000,000 Bq
Exposure (*X*)	röntgen	R	esu / 0.001293 g of air	1928	2.58×10^{-4} C/kg
Fluence (Φ)	(reciprocal area)		m^{-2}	1962	SI
Absorbed dose (*D*)	erg		$erg{\cdot}g^{-1}$	1950	1.0×10^{-4} Gy
	rad	rad	$100\ erg{\cdot}g^{-1}$	1953	0.010 Gy
	gray	Gy	$J{\cdot}kg^{-1}$	1974	SI
Dose equivalent (*H*)	röntgen equivalent man	rem	$100\ erg{\cdot}g^{-1}$	1971	0.010 Sv
	sievert	Sv	$J{\cdot}kg^{-1} \times W_R$	1977	SI

The measure of the ionizing effect of gamma and X-rays in dry air is called the exposure, for which a legacy unit, the röntgen was used from 1928. This has been replaced by kerma, now mainly used for instrument calibration purposes but not for received dose effect. The effect of gamma and other ionizing radiation on living tissue is more closely related to the amount of energy deposited in tissue rather than the ionisation of air, and replacement radiometric units and quantities for radiation protection have been defined and developed from 1953 onwards. These are:

- The gray (Gy), is the SI unit of absorbed dose, which is the amount of radiation energy deposited in the irradiated material. For gamma radiation this is numerically equivalent to equivalent dose measured by the sievert, which indicates the stochastic biological effect of low levels of radiation on human tissue. The radiation weighting conversion factor from absorbed dose to equivalent dose is 1 for gamma, whereas alpha particles have a factor of 20, reflecting their greater ionising effect on tissue.
- The rad is the deprecated CGS unit for absorbed dose and the rem is the deprecated CGS unit of equivalent dose, used mainly in the USA.

Distinction from X-rays

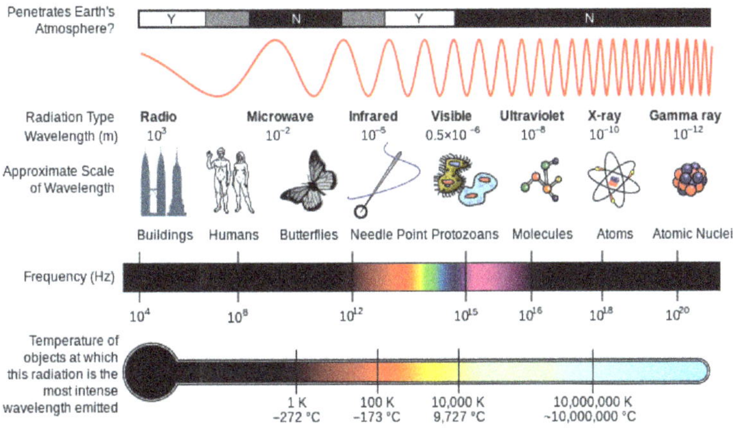

In practice, gamma ray energies overlap with the range of X-rays, especially in the higher-frequency region referred to as "hard" X-rays. This depiction follows the older convention of distinguishing by wavelength.

The distinction between X-rays and gamma rays has changed in recent decades. Originally, the electromagnetic radiation emitted by X-ray tubes almost invariably had a longer wavelength than the radiation (gamma rays) emitted by radioactive nuclei. Older literature distinguished between X- and gamma radiation on the basis of wavelength, with radiation shorter than some arbitrary wavelength, such as 10^{-11} m, defined as gamma rays. Since the energy of photons is proportional to their frequency and inversely proportional to wavelength, this past distinction between X-rays and gamma rays can also be thought of in terms of its energy, with gamma rays considered to be higher energy electromagnetic radiation than are X-rays.

However, current artificial sources now able to duplicate any electromagnetic radiation that originates in the nucleus, as well as far higher energies, the wavelengths characteristic of radioactive gamma ray sources vs. other types now completely overlap. Thus, gamma rays are now usually distinguished by their origin: X-rays are emitted by definition by electrons outside the nucleus, while gamma rays are emitted by the nucleus. Exceptions to this convention occur in astronomy, where gamma decay is seen in the afterglow of certain supernovas, but radiation from high energy processes known to involve other radiation sources than radioactive decay is still classed as gamma radiation.

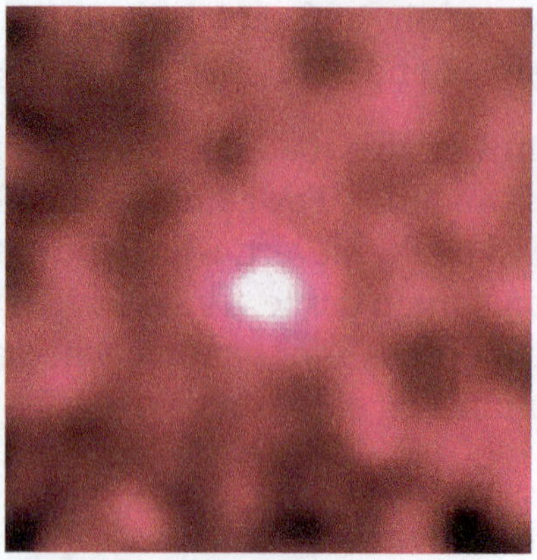

The Moon as seen by the Compton Gamma Ray Observatory, in gamma rays of greater than 20 MeV. These are produced by cosmic ray bombardment of its surface. The Sun, which has no similar surface of high atomic number to act as target for cosmic rays, cannot usually be seen at all at these energies, which are too high to emerge from primary nuclear reactions, such as solar nuclear fusion (though occasionally the Sun produces gamma rays by cyclotron-type mechanisms, during solar flares).Gamma rays have higher energy than X-rays.

For example, modern high-energy X-rays produced by linear accelerators for megavoltage treatment in cancer often have higher energy (4 to 25 MeV) than do most classical gamma rays produced by nuclear gamma decay. One of the most common gamma ray emitting isotopes used in diagnostic nuclear medicine, technetium-99m, produces gamma radiation of the same energy (140 keV) as that produced by diagnostic X-ray machines, but of significantly lower energy than therapeutic photons from linear particle accelerators. In the medical community today, the convention that radiation produced by nuclear decay is the only type referred to as "gamma" radiation is still respected.

Due to this broad overlap in energy ranges, in physics the two types of electromagnetic radiation are now often defined by their origin: X-rays are emitted by electrons (either in orbitals outside of the nucleus, or while being accelerated to produce bremsstrahlung-type radiation), while gamma rays are emitted by the nucleus or by means of other particle decays or annihilation events. There is no lower limit to the energy of photons produced by nuclear reactions, and thus ultraviolet or lower energy photons produced by these processes would also be defined as "gamma rays". The only naming-convention that is still universally respected is the rule that electromagnetic radiation that is known to be of atomic nuclear origin is *always* referred to as "gamma rays", and never as X-rays. However, in physics and astronomy, the converse convention (that all gamma rays are considered to be of nuclear origin) is frequently violated.

In astronomy, higher energy gamma and X-rays are defined by energy, since the processes that produce them may be uncertain and photon energy, not origin, determines the required astronomical detectors needed. High-energy photons occur in nature that are known to be produced by processes other than nuclear decay but are still referred to as gamma radiation. An example is "gamma rays" from lightning discharges at 10 to 20 MeV, and known to be produced by the bremsstrahlung mechanism.

Another example is gamma-ray bursts, now known to be produced from processes too powerful to involve simple collections of atoms undergoing radioactive decay. This has led to the realization that many gamma rays produced in astronomical processes result not from radioactive decay or particle annihilation, but rather in much the same manner as the production of X-rays. Although gamma rays in astronomy are discussed below as non-radioactive events, in fact a few gamma rays are known in astronomy to originate explicitly from gamma decay of nuclei (as demonstrated by their spectra and emission half life). A classic example is that of supernova SN 1987A, which emits an "afterglow" of gamma-ray photons from the decay of newly made radioactive nickel-56 and cobalt-56. Most gamma rays in astronomy, however, arise by other mechanisms.

Law of Radioactivity Decay

When a radioactive material undergoes α, β or γ-decay, the number of nuclei undergoing the decay, per unit time, is proportional to the total number of nuclei in the sample material. So,

If N = total number of nuclei in the sample and ΔN = number of nuclei that undergo decay in time Δt then,

$$\Delta N / \Delta t \propto N$$

Or, $\Delta N / \Delta t = \lambda N$

where λ = radioactive decay constant or disintegration constant. Now, the change in the number of nuclei in the sample is, dN = – ΔN in time Δt. Hence, the rate of change of N (in the limit Δt→ 0) is,

$$\mathrm{dN} / \mathrm{dt} = -\lambda \mathrm{N}$$

Or, $dN / N = -\lambda dt$

Now, integrating both sides of the above equation, we get,

$$^{N}{}_{N0}\int dN / N = \lambda^{t}{}_{t0}\int dt$$

Or, $\ln N - ln\, N_0 = -\lambda\,(t - t_0)$

Where, N_0 is the number of radioactive nuclei in the sample at some arbitrary time $t_0$and N is the number of radioactive nuclei at any subsequent time t. Next, we set t_0 = 0 and rearrange the above equation to get,

$\ln(N / N_0) = -\lambda t$

Or, $N(t) = N_0 e^{-\lambda t}$

Above Equation is the Law of Radioactive Decay.

The Decay Rate

In radioactivity calculations, we are more interested in the decay rate R (= – dN/dt) than in N itself. This rate gives us the number of nuclei decaying per unit time. Even if we don't know the number of nuclei in the sample, by simply measuring the number of emissions of α, β or γ particles in 10 or 20 seconds, we can calculate the decay rate. Let's say that we consider a time interval dt and get a decay count ΔN (= $-dN$). The Decay rate is now defined as,

$$R = -dN/dt$$

Differentiating equation $N(t) = N_0 e^{-\lambda t}$ on both sides, we get,

$$R = \lambda N_0 e^{-\lambda t}$$

Or, $R = R_0 e^{-\lambda t}$

Where, R_0 is the radioactive decay rate at time t = 0, and R is the rate at any subsequent time t. Above equation is the alternative form of the law of radioactive decay. Now we can rewrite equation $\Delta N / \Delta t = \lambda N$ as follows,

$$R = \lambda N$$

where R and the number of radioactive nuclei that have not yet undergone decay must be evaluated at the same instant.

Half-Life and Mean Life

The total decay rate of a sample is also known as the *activity* of the sample. The SI unit for measurement of activity is 'becquerel' and is defined as,

1 becquerel = 1Bq = 1 decay per second

An older unit, the curie, is still in common use:

$$1\ curie = 1\ Ci = 3.7 \times 10^{10}\ Bq\ (decays\ per\ second)$$

There are two ways to measure the time for which a radionuclide can last.

- Half-life $T_{1/2}$ – the time at which both R and N are reduced to half of their initial values
- Mean life τ – the time at which both R and N have been reduced to, e^{-1} of their initial values.

Calculating Half-Life

Let's find the relation between T1/2 and the disintegration constant λ. For this, let's input the following values in equation $R = R_0 e^{-\lambda t}$,

$R = (1/2)R_0$ and $t = T_{1/2}$

So, we get $T_{1/2} = (ln2)/\lambda$

Or, $T_{1/2} = 0.693/\lambda$

Calculating Mean Life

Next, let's find the relation between the mean life τ and the disintegration constant λ. For this, let's consider equation $R = R_0 e^{-\lambda t}$,

- The number of nuclei which decay in the time interval: 't' to '$t + \Delta t$' is:

  $R(t)\Delta t = (\lambda N_0 e^{-\lambda t} \Delta t)$.
- Each of them has lived for time't'.
- Hence, the total life of all these nuclei is $t\lambda N_0 e^{-\lambda t} \Delta t$

Hence, to obtain the mean life, we integrate this expression over all the times from 0 to ∞ and divide by the total number of nuclei at t = 0 (which is N_0).

$$\tau = (\lambda N_0 {}^{0}{}_{\infty} \int te^{-\lambda t} dt) / N_0$$
$$= \lambda {}^{0}{}_{\infty} \int te^{-\lambda t} dt$$

On solving this integral, we get

$$\tau = 1/\lambda$$

Therefore, we can summarise the observations as follows:

$$T1/2 = (ln2)/\lambda = \tau \, ln\, 2$$

Example: The half-life of ${}^{238}{}_{92}U$ undergoing α-decay is 4.5×10^9 years. What is the activity of 1g sample of ${}^{238}{}_{92}U$?

Solution: T1/2 = 4.5 x 10^9 years = 4.5 x 10^9 years x 3.6 x 10^7 seconds/year = 1.42 x 10^{17}seconds

We know that 1 k mol of any isotope contains Avogadro's number of atoms. Hence, 1g of ${}^{238}{}_{92}U$ contains, {1/(238 x 10^{-3})} x 6.025 x 10^{26} = 25.3 x 10^{20} atoms. Therefore, the decay rate R is,

$$R = \lambda N$$
$$= (0.693 / T_{1/2})N$$
$$= (0.693\, x\, 25.3\, x\, 10^{20}) \;/\; (1.42\, x\, 10^{17})$$

Therefore, $R = 1.23\, x\, 10^4\, Bq$

Chain of Two Decays

Now consider the case of a chain of two decays: one nuclide *A* decaying into another *B* by one process, then *B* decaying into another *C* by a second process, i.e. $A \rightarrow B \rightarrow C$. The previous equation cannot be applied to the decay chain, but can be generalized as follows. Since *A* decays into *B*, *then B* decays into *C*, the activity of *A* adds to the total number of *B* nuclides in the present sample, *before* those *B* nuclides decay and reduce the number of nuclides leading to the later sample. In other words, the number of second generation nuclei *B* increases as a result of the first generation nuclei decay of *A*, and decreases as a result of its own decay into the third generation nuclei *C*. The sum of these two terms gives the law for a decay chain for two nuclides:

$$\frac{dN_B}{dt} = -\lambda_B N_B + \lambda_A N_A.$$

The rate of change of N_B, that is dN_B/dt, is related to the changes in the amounts of *A* and *B*, N_B can increase as *B* is produced from *A* and decrease as *B* produces *C*.

Re-writing using the previous results:

$$\frac{dN_B}{dt} = -\lambda_B N_B + \lambda_A N_{Ao} e^{-\lambda_A t}$$

The subscripts simply refer to the respective nuclides, i.e. N_A is the number of nuclides of type *A*, N_{Ao} is the initial number of nuclides of type *A*, λ_A is the decay constant for *A* - and similarly for nuclide *B*. Solving this equation for N_B gives:

$$N_B = \frac{N_{Ao}\lambda_A}{\lambda_B - \lambda_A}\left(e^{-\lambda_A t} - e^{-\lambda_B t}\right).$$

In the case where *B* is a stable nuclide ($\lambda_B = 0$), this equation reduces to the previous solution:

$$\lim_{\lambda_B \to 0}\left[\frac{N_{Ao}\lambda_A}{\lambda_B - \lambda_A}\left(e^{-\lambda_A t} - e^{-\lambda_B t}\right)\right] = \frac{N_{Ao}\lambda_A}{0 - \lambda_A}\left(e^{-\lambda_A t} - 1\right) = N_{Ao}\left(1 - e^{-\lambda_A t}\right),$$

as shown above for one decay. The solution can be found by the integration factor method, where the integrating factor is $e^{\lambda_B t}$. This case is perhaps the most useful, since it can derive both the one-decay equation (above) and the equation for multi-decay chains (below) more directly.

Chain of any Number of Decays

For the general case of any number of consecutive decays in a decay chain, i.e. $A_1 \rightarrow A_2 \cdots \rightarrow A_i \cdots \rightarrow A_D$, where *D* is the number of decays and *i* is a dummy index ($i = {}_{1, 2, 3, \ldots D}$), each nuclide population can be found in terms of the previous population. In this case $N_2 = 0$, $N_3 = 0, \ldots, N_D = 0$. Using the above result in a recursive form:

$$\frac{dN_j}{dt} = -\lambda_j N_j + \lambda_{j-1} N_{(j-1)o} e^{-\lambda_{j-1} t}.$$

The general solution to the recursive problem is given by *Bateman's equations*:

Bateman's equations

$$ü_{ü} = \frac{N_1 ü}{\lambda_D} \sum_{i=1}^{D} \lambda \quad {}^{-\lambda_i t}$$

$$c_i = \prod_{j=1, i \neq j}^{D} \frac{\lambda_j}{\lambda_j - \lambda_i}$$

Alternative Decay Modes

In all of the above examples, the initial nuclide decays into just one product. Consider the case of one initial nuclide that can decay into either of two products, that is $A \to B$ and $A \to C$ in parallel. For example, in a sample of potassium-40, 89.3% of the nuclei decay to calcium-40 and 10.7% to argon-40. We have for all time *t*:

$$N = N_A + N_B + N_C$$

which is constant, since the total number of nuclides remains constant. Differentiating with respect to time:

$$\frac{\mathrm{d}N_A}{\mathrm{d}t} = -\left(\frac{\mathrm{d}N_B}{\mathrm{d}t} + \frac{\mathrm{d}N_C}{\mathrm{d}t}\right)$$

$$-\lambda N_A = -N_A\left(\lambda_B + \lambda_C\right)$$

defining the *total decay constant* λ in terms of the sum of *partial decay constants* λ_B and λ_C:

$$\lambda = \lambda_B + \lambda_C.$$

Notice that

$$\frac{\mathrm{d}N_A}{\mathrm{d}t} < 0, \frac{\mathrm{d}N_B}{\mathrm{d}t} > 0, \frac{\mathrm{d}N_C}{\mathrm{d}t} > 0.$$

Solving this equation for N_A:

$$\ddot{u}_{A} = {}_{A0}{}^{-\lambda t}.$$

where N_{A0} is the initial number of nuclide A. When measuring the production of one nuclide, one can only observe the total decay constant λ. The decay constants λ_B and λ_C determine the probability for the decay to result in products *B* or *C* as follows:

$$N_B = \frac{\lambda_B}{\lambda} N_{A0}\left(1 - e^{-\lambda t}\right),$$

$$N_C = \frac{\lambda_C}{\lambda} N_{A0}\left(1 - e^{-\lambda t}\right).$$

because the fraction λ_B/λ of nuclei decay into *B* while the fraction λ_C/λ of nuclei decay into *C*.

Corollaries of the Decay Laws

The above equations can also be written using quantities related to the number of nuclide particles *N* in a sample;

- The activity: $A = \lambda N$.

- The amount of substance: $n = N/L$.
- The mass: $M = A_r n = A_r N/L$.

where $L = 6.022\times10^{23}$ is Avogadro's constant, A_r is the relative atomic mass number, and the amount of the substance is in moles.

Example – Radioactive Decay Law

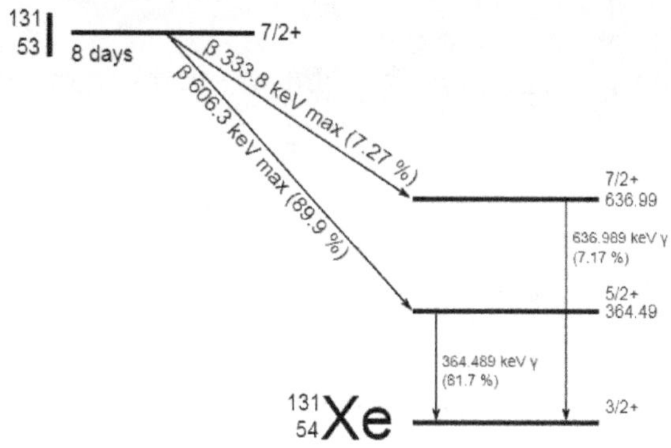

A sample of material contains 1 mikrogram of iodine-131. Note that, iodine-131 plays a major role as a radioactive isotope present in nuclear fission products, and it a major contributor to the health hazards when released into the atmosphere during an accident. Iodine-131 has a half-life of 8.02 days.

Calculate:

1. The number of iodine-131 atoms initially present.
2. The activity of the iodine-131 in curies.
3. The number of iodine-131 atoms that will remain in 50 days.
4. The time it will take for the activity to reach 0.1 mCi.

Solution:

1. The number of atoms of iodine-131 can be determined using isotopic mass as below.

$$N_{I-131} = m_{I-131} \cdot N_A \;/\; M_{I-131}$$
$$N_{I-131} = (1\ \mu g) \times (6.02\times10^{23}\ nuclei/mol) \;/\; (130.91\ g/mol)$$
$$N_{I-131} = 4.6 \times 10^{15}\ nuclei$$

2. The activity of the iodine-131 in curies can be determined using its decay constant:

The iodine-131 has half-live of 8.02 days (692928 sec) and therefore its decay constant is:

$$\lambda\left(s^{-1}\right) = \frac{\ln 2}{t_{1/2}}$$

$$\lambda\left(s^{-1}\right) = \frac{0.693}{692928} = 1\,x\,10^{-6}\,s^{-1}$$

Using this value for the decay constant we can determine the activity of the sample:

$$A = N\ x\ \lambda = 4.6\ x\ 10^{15}\ x\ 1\ x\ 10^{-6} = 4.6\ x\ 10^{9}\ Bq = 0.124\ Ci$$

3) and 4) The number of iodine-131 atoms that will remain in 50 days (N_{50d}) and the time it will take for the activity to reach 0.1 mCi can be calculated using the decay law:

3)

$$A = N\ x\ \lambda = 4.6\ x\ 10^{15}\ x\ 1\ x\ 10^{-6} = 4.6\ x\ 10^{9}\ Bq = 0.124\ Ci$$

$$N_{50d} = N.e^{-\lambda.t} = 4.6\ 10^{15}\ .\ e^{-1x10^{-6}.4320000} = 6.12\ x\ 10^{13}$$

4)

$$A = A_0.e^{-\lambda.t}$$

$$t_{0.1mCi} = \frac{-ln\frac{A}{A_0}}{\lambda} = \frac{-ln\frac{0.1mCi}{0.124Ci}}{1\ x\ 10^{-6}} = 82\ days$$

As can be seen, after 50 days the number of iodine-131 atoms and thus the activity will be about 75 times lower. After 82 days the activity will be approximately 1200 times lower. Therefore, the time of ten half-lives (factor 2^{10}= 1024) is widely used to define residual activity.

References

- Woodside, Gayle (1997). Environmental, Safety, and Health Engineering. US: John Wiley & Sons. p. 476. ISBN 0471109320. Archived from the original on 2015-10-19
- Radioactivity, nuclear-chemistry: tutorvista.com, Retrieved 28 April 2018
- Liebel F, Kaur S, Ruvolo E, Kollias N, Southall MD (2012). "Irradiation of skin with visible light induces reactive oxygen species and matrix-degrading enzymes". J. Invest. Dermatol. 132 (7): 1901–1907. doi:10.1038/jid.2011.476. PMID 22318388
- Radionuclide: nuclear-energy.net, Retrieved 18 July 2018
- W.-M. Yao; et al. (2007). "Particle Data Group Summary Data Table on Baryons" (PDF). J. Phys. G. 33 (1). Archived from the original (PDF) on 2011-09-10. Retrieved 2012-08-16
- Ionizing-radiation: techtarget.com, Retrieved 26 March 2018
- Pattison JE, Bachmann DJ, Beddoe AH (1996). "Gamma Dosimetry at Surfaces of Cylindrical Containers". Journal of Radiological Protection. 16 (4): 249–261. Bibcode:1996JRP....16..249P. doi:10.1088/0952-4746/16/4/004.
- Gamma-ray, science: britannica.com, Retrieved 20 May 2018
- Stallcup, James G. (2006). OSHA: Stallcup's High-voltage Telecommunications Regulations Simplified. US: Jones & Bartlett Learning. p. 133. ISBN 076374347X. Archived from the original on 2015-10-17.
- Radioactivity-law-of-radioactive-decay, physics: toppr.com, Retrieved 10 July 2018
- Mortazavi, S.M.J.; P.A. Karamb (2005). "Apparent lack of radiation susceptibility among residents of the high background radiation area in Ramsar, Iran: can we relax our standards?". Radioactivity in the Environment. 7: 1141–1147. doi:10.1016/S1569-4860(04)07140-2. ISSN 1569-4860.

3

Nuclear Energetics

An understanding of nuclear energetics follows from an in-depth study of nuclear energy, mass energy balance, binding energy, Q-value and nuclide stability. This chapter has been written to elucidate the fundamentals of these key concepts in a comprehensive language.

Nuclear Energy

Nuclear energy is energy in the nucleus (core) of an atom. Atoms are tiny particles that make up every object in the universe. There is enormous energy in the bonds that hold atoms together. Nuclear energy can be used to make electricity. But first the energy must be released. It can be released from atoms in two ways: nuclear fusion and nuclear fission. In nuclear fusion, energy is released when atoms are combined or fused together to form a larger atom. This is how the sun produces energy. In nuclear fission, atoms are split apart to form smaller atoms, releasing energy. Nuclear power plants use nuclear fission to produce electricity.

Fission and Fusion

There are two fundamental nuclear processes considered for energy production: fission and fusion.

- Fission is the energetic splitting of large atoms such as Uranium or Plutonium into two smaller atoms, called fission products. To split an atom, you have to hit it with a neutron. Several neutrons are also released which can go on to split other nearby atoms, producing a nuclear chain reaction of sustained energy release. This nuclear reaction was the first of the two to be discovered. All commercial nuclear power plants in operation use this reaction to generate heat which they turn into electricity.
- Fusion is the combining of two small atoms such as Hydrogen or Helium to produce heavier atoms and energy. These reactions can release more energy than fission without producing as many radioactive byproducts. Fusion reactions occur in the sun, generally using Hydrogen as fuel and producing Helium as waste (Helium was discovered in the sun and named after the Greek Sun God, Helios). This reaction has not been commercially developed yet and is a serious research interest worldwide, due to its promise of nearly limitless, low-pollution, and non-proliferative energy.

Energy Density of Various Fuel Sources

The amount of energy released in nuclear reactions is astounding. Table shows how long a 100 Watt light bulb could run from using 1 kg of various fuels. The natural uranium undergoes nuclear fission and thus attains very high energy density (energy stored in a unit of mass).

Material	Energy Density (MJ/kg)	100W light bulb time (1kg)
Wood	10	1.2 days
Ethanol	26.8	3.1 days
Coal	32.5	3.8 days
Crude oil	41.9	4.8 days
Diesel	45.8	5.3 days
Natural Uranium (LWR)	$5.7x10^5$	182 years
Reactor Grade Uranium (LWR)	$3.7x10^6$	1,171 years
Natural Uranium (breeder)	$8.1x10^7$	25,700 years
Thorium (breeder)	$7.9x10^7$	25,300 years

Table above energy densities of various energy sources in MJ/kg and in length of time that 1 kg of each material could run a 100W load. Natural uranium has undergone no enrichment (0.7% U-235), reactor-grade uranium has 5% U-235. By the way, 1 kg of weapons grade uranium (95% U-235) could power the entire USA for 177 seconds. All numbers assume 100% thermal-to-electrical conversion.

Capabilities of Nuclear Power

Sustainable

Table sums the sustainability of nuclear power up quite well. However, there is quite a bit of talk about nuclear fuel (Uranium) running low just like oil. Technically, this is a non-issue, as nuclear waste is recyclable. Economically, it could become a major issue. Today's commercial nuclear reactors burn less than 1% of the fuel that is mined for them and the rest of it or so is thrown away (as depleted uranium and nuclear waste). The US recycling program shut down in the '70s due to proliferation and economic concerns. Today, France and Japan are recycling fuel with great success. New technology exists that can greatly reduce proliferation concerns. IAEA suggests that there are over 200 years of Uranium reserves at current demand. There is also a very large supply of uranium dissolved in seawater at very low concentration. No one has found a cheap-enough way to extract it yet, though people have come close. Nuclear reactors can also run on Thorium fuel.

Ecological

In operation, nuclear power plants emit nothing into the environment except hot water. The classic cooling tower icon of nuclear reactors is just that, a cooling tower. Clean water vapor is all that comes out. Very little CO_2 or other climate-changing gases come out of nuclear power generation (certainly some CO_2 is produced during mining, construction, etc., but the amount is about 50 times less than coal and 25 times less than natural gas plants.). The spent nuclear fuel (nuclear waste) can be handled properly and disposed of geologically without affecting the environment in any way.

Independent

With nuclear power, many countries can approach energy independence. Being "addicted to oil" is a major national and global security concern for various reasons. Using electric or plug-in hybrid electric vehicles (PHEVs)powered by nuclear reactors, we could reduce our oil demands by orders of magnitude. Additionally, many nuclear reactor designs can provide high-quality process heat in addition to electricity, which can in turn be used to desalinate water, prepare hydrogen for fuel cells, or to heat neighborhoods, among many other industrial processes.

Nuclear Energetics

Nuclear Energetics is the energy change in nuclear reactions is due to the gain or loss of mass in the reaction.

The energy stored in atomic nuclei is more than a million times greater than that from Chemical reactions and is a driving force in the evolution of our Universe. The energy radiated by our Sun is the consequence of nuclear reactions that occur in its core. One of the greatest hopes for clean, abundant energy in the future is in the nuclear fusion reactor, which utilizes similar reactions to produce electrical energy. The elements that make up our terrestial environment are the products of nuclear reactions generated during the phases of Stellar Evolution. One of these elements, 235U, is the fuel for nuclear power reactors. And miniature power sources used in many remote-sensing devices operate with the energy provided by radioactive decay.

Not all combinations of protons and neutrons are able to form a unique nucleus, just as all combinations of atoms do not necessarily form stable compounds; e.g. two He atoms do not form He_2.

Basics

Nuclear energetic calculations are simplified relative to chemical thermodynamics in that for most applications, the entropy is zero (stellar interiors being a notable exception). The basic thermodynamic equation is

$E = Mc^2$ (with thanks to A. Einstein).

Mass M is expressed in units of atomic mass units u (or amu) and energy E in terms of MeV. In this context, the value of c^2 is

$$c^2 = 931.494\, MeV / u.$$

Exercise: using the definition of u in grams and MeV in ergs, show that 1u = 931.494 MeV.

Nuclear energetics, mass and energy can be treated interchangeably, either in units of u or MeV/c^2, as illustrated in table.

Symbol	Description	Mass in grams	Mass in u	Mass in MeV/c^2
Mp	mass of proton	1.6726×10^{-24}	1.00728	938.272
Me	mass of electron	9.1×10^{-28}	5.484×10^{-4}	0.511006
MH	mass of 1H atom	1.6735×10^{-24}	1.00783	938.783
Mn	mass of neutron	1.6748×10^{-24}	1.00865	939.550

Two features of table should be noted. First, the masses of the proton and neutron are ~1830-1840 times larger than the mass of the electron, indicating the mass of the atom is concentrated in its nucleus. And second the mass of the neutron is greater than that of the proton, which has important implications for our Universe.

In order to perform nuclear energetic calculations, it is more convenient to work in units of MeV, rather than mass. In order to simplify calculations, we define a quantity called the mass defect Δ (sometimes called the mass excess):

$$(M - A)c^2 \text{ (units are in MeV)}$$

or the mass can be calculated from

$$\ddot{u} \ \varnothing \qquad / \ ^2 \text{ (units are in u)}$$

Exercise: Calculate the mass defect for the ^{4}He atom, given M (^{4}He) = 4.002603 u.

$$D = (4.002603 - 4)u\ (931.494 MeV / u) = 2.425\ MeV$$

It is important to stress that the chemical atomic weight listed in the Periodic table is Not the same as the mass of an individual isotope, unless the element has only one stable isotope; e.g. ^{27}Al. The chemical atomic weight is the average of all stable isotopes of a given element, $< M_Z >$,

$$< M_Z > = \Sigma f_i M_i,$$

where f^i is the relative abundance of each isotope of element Z (given in the Nuclear

Wallet Cards) and Mi is the mass of each isotope ($M = A + \Delta / c^2$).

Example: Calculate the chemical atomic weight of copper, element 29.

f (^{63}Cu) = 69.09%; M(^{63}Cu) = 62.92959u
f(^{65}Cu) = 30.91%; M(^{65}Cu) = 64.92779u
<MCu> = (0.6909)(62.92959u) + (0.3091)(64.92779u) = 63.54u

Relativistic effects may be important for very high velocity particles; i.e. the total mass of a particle increases with its velocity, v. For example, a 200-MeV proton (rest mass Mo) accelerated at the IU Cyclotron Facility has a velocity of 0.6c, which increases its total mass by 25%, or $M / M_0 = 1.25$. We will use the classical equations of motion and assume relativity is negligible, except for a few special cases,

Kinetic energy: $EK = Mov^2 / 2$

Linear momentum: $p = M_0 v$

These relationships are accurate to 1% for v/c < 0.1.

Mass Energy Balance

The relationship between mass (m) and energy (E) is expressed in the following equation:

$$E = mc^2$$

where

- c is the speed of light ($2.998 \times 10^8 m / s$), and
- E and m are expressed in units of joules and kilograms, respectively.

Albert Einstein first derived this relationship in 1905 as part of his special theory of relativity: the mass of a particle is directly proportional to its energy. Thus according to equation $E = mc^2$, every mass has an associated energy, and similarly, any reaction that involves a change in energy must be accompanied by a change in mass. This implies that all exothermic reactions should be accompanied by a decrease in mass, and all endothermic reactions should be accompanied by an increase in mass. Given the law of conservation of mass, how can this be true? The solution to this apparent contradiction is that chemical reactions are indeed accompanied by changes in mass, but these changes are simply too small to be detected. As you may recall, all particles exhibit wavelike behavior, but the wavelength is inversely proportional to the mass of the particle (actually, to its momentum, the product of its mass and velocity). Consequently, wavelike behavior is detectable only for particles with very small masses, such as electrons. For example, the chemical equation for the combustion of graphite to produce carbon dioxide is as follows:

$$C(\text{graphite}) + \frac{1}{2} O_2(g) \rightarrow CO_2(g) \quad \Delta H^\circ = -393.5 \text{ kJ/mol}$$

Combustion reactions are typically carried out at constant pressure, and under these conditions, the heat released or absorbed is equal to ΔH. When a reaction is carried out at constant volume, the heat released or absorbed is equal to ΔE. For most chemical reactions, however, $\Delta E \approx \Delta H$. If we rewrite Einstein's equation as:

$$\Delta E = (\Delta m)c^2$$

we can rearrange the equation to obtain the following relationship between the change in mass and the change in energy:

$$\Delta m = \frac{\Delta E}{c^2}$$

Because $1\,J = 1\,(kg \bullet m^2)/s^2$, the change in mass is as follows:

$$\Delta m = \frac{-393.5 \text{ kJ/mol}}{(2.998\times10^8 \text{ m/s})^2} = \frac{-3.935\times105(\text{kg}\times\text{m}^2)/(\text{s}^2\times\text{mol})}{(2.99^8\times10^8 \text{ m/s})^2} = -4.38\times10^{-12} \text{ kg/mol}$$

This is a mass change of about 3.6×10^{-10} g/g carbon that is burned, or about 100-millionths of the mass of an electron per atom of carbon. In practice, this mass change is much too small to be measured experimentally and is negligible.

In contrast, for a typical nuclear reaction, such as the radioactive decay of ^{14}C to ^{14}N and an electron (a β particle), there is a much larger change in mass:

$$^{14}\text{C} \rightarrow {}^{14}\text{N} + {}^{0}_{-1}\beta$$

We can use the experimentally measured masses of subatomic particles and common isotopes given in table to calculate the change in mass directly. The reaction involves the conversion of a neutral ^{14}C atom to a positively charged ^{14}N ion (with six, not seven, electrons) and a negatively charged β particle (an electron), so the mass of the products is identical to the mass of a neutral ^{14}N atom. The total change in mass during the reaction is therefore the difference between the mass of a neutral ^{14}N atom (14.003074 amu) and the mass of a ^{14}C atom (14.003242 amu):

$$\begin{aligned}\Delta m &= \text{mass}_{\text{products}} - \text{mass}_{\text{reactants}}\\ &= 14.003074 \text{ amu} - 14.003242 \text{ amu} = -0.000168 \text{ amu}\end{aligned}$$

The difference in mass, which has been released as energy, corresponds to almost one-third of an electron. The change in mass for the decay of 1 mol of ^{14}C is -0.000168 g = -1.68×10^{-4}g = -1.68×10^{-7} kg. Although a mass change of this magnitude may seem small, it is about 1000 times larger than the mass change for the combustion of graphite. The energy change is as follows:

$$\begin{aligned}\Delta E &= (\Delta m)c^2 = (-1.68\times10^{-7} \text{ kg})(2.998\times10^8 \text{ m/s})^2\\ &= -1.51\times10^{10}(\text{kg}\times\text{m}^2)/\text{s}^2 = -1.51\times10^{10}\,J = -1.51\times10^7 \text{ kJ}\end{aligned}$$

The energy released in this nuclear reaction is more than 100,000 times greater than that of a typical chemical reaction, even though the decay of ^{14}C is a relatively low-energy nuclear reaction.

Because the energy changes in nuclear reactions are so large, they are often expressed in kiloelectronvolts ($1\,keV = 10^3\,eV$), megaelectronvolts ($1\,MeV = 10^6\,eV$), and even gigaelectronvolts ($1\,GeV = 10^9\,eV$) per atom or particle. The change in energy that accompanies a nuclear reaction can be calculated from the change in mass using the relationship 1 amu = 931 MeV. The energy released by the decay of one atom of ^{14}C is thus

$$\left(-1.68\times10^{-4}\,\text{amu}\right)\left(\frac{931\,\text{MeV}}{\text{amu}}\right)=-0.156\,\text{MeV}=-156\,\text{keV}$$

Example

Calculate the changes in mass (in atomic mass units) and energy (in joules per mole and electronvolts per atom) that accompany the radioactive decay of ^{238}U to ^{234}Th and an α particle. The α particle absorbs two electrons from the surrounding matter to form a helium atom.

Given: nuclear decay reaction

Asked for: changes in mass and energy

Strategy

A Use the mass values in table to calculate the change in mass for the decay reaction in atomic mass units.

B Use Equation $\Delta m=\frac{\Delta E}{c^2}$ to calculate the change in energy in joules per mole.

C Use the relationship between atomic mass units and megaelectronvolts to calculate the change in energy in electronvolts per atom.

Solution

A Using particle and isotope masses from table, we can calculate the change in mass as follows:

$$\Delta m=\text{mass}_{\text{products}}-\text{mass}_{\text{reactants}}=(\text{mass}^{234}\,Th+\text{mass}\,^{4}_{2}\text{He})-\text{mass}^{238}\,\text{U}$$

$$=(234.043601\text{ amu}+4.002603\text{ amu})-238.050788\text{ amu}=-0.004584\text{ amu}$$

B Thus the change in mass for 1 mol of ^{238}U is −0.004584 g or -4.584×10^{-6} kg. The change in energy in joules per mole is as follows:

$$\Delta E=(\Delta m)c^2=(-4.584\times10^{-6}\,kg)(2.998\times10^{8}\,m/s)^2=-4.120\times10^{11}\,J/mol$$

C The change in energy in electronvolts per atom is as follows:

$$\Delta E=-4.584\times10^{-3}\text{ amu}\times\frac{931\,\text{MeV}}{\text{amu}}\times\frac{1\times10^{6}\text{ eV}}{1\,\text{MeV}}=-4.27\times10^{6}\text{ eV/ atom}$$

Exercise

Calculate the changes in mass (in atomic mass units) and energy (in kilojoules per mole and kiloelectronvolts per atom) that accompany the radioactive decay of tritium (^{3}H) to ^{3}He and a β particle.

$$\Delta m=-2.0\times10^{-5}\,amu;\ \Delta E=-1.9\times10^{6}\,kJ/mol=-19\,keV/atom$$

Conservation of Mass and Energy

Mass and energy can be seen as two names (and two measurement units) for the same underlying, conserved physical quantity. Thus, the laws of conservation of energy and conservation of (total) mass are equivalent and both hold true. Einstein elaborated in a 1946 essay that "the principle of the conservation of mass proved inadequate in the face of the special theory of relativity. It was therefore merged with the energy conservation principle—just as, about 60 years before, the principle of the conservation of mechanical energy had been combined with the principle of the conservation of heat [thermal energy]. We might say that the principle of the conservation of energy, having previously swallowed up that of the conservation of heat, now proceeded to swallow that of the conservation of mass—and holds the field alone."

If the conservation of mass law is interpreted as conservation of *rest* mass, it does not hold true in special relativity. The *rest* energy (equivalently, rest mass) of a particle can be converted, not "to energy" (it already *is* energy (mass)), but rather to *other* forms of energy (mass) that require motion, such as kinetic energy, thermal energy, or radiant energy. Similarly, kinetic or radiant energy can be converted to other kinds of particles that have rest energy (rest mass). In the transformation process, neither the total amount of mass nor the total amount of energy changes, since both properties are connected via a simple constant. This view requires that if either energy or (total) mass disappears from a system, it is always found that both have simply moved to another place, where they are both measurable as an increase of both energy and mass that corresponds to the loss in the first system.

Fast-moving Objects and Systems of Objects

When an object is pushed in the direction of motion, it gains momentum and energy, but when the object is already traveling near the speed of light, it cannot move much faster, no matter how much energy it absorbs. Its momentum and energy continue to increase without bounds, whereas its speed approaches (but never reaches) a constant value—the speed of light. This implies that in relativity the momentum of an object cannot be a constant times the velocity, nor can the kinetic energy be a constant times the square of the velocity.

A property called the relativistic mass is defined as the ratio of the momentum of an object to its velocity. Relativistic mass depends on the motion of the object, so that different observers in relative motion see different values for it. If the object is moving slowly, the relativistic mass is nearly equal to the rest mass and both are nearly equal to the usual Newtonian mass. If the object is moving quickly, the relativistic mass is greater than the rest mass by an amount equal to the mass associated with the kinetic energy of the object. As the object approaches the speed of light, the relativistic mass grows infinitely, because the kinetic energy grows infinitely and this energy is associated with mass.

The relativistic mass is always equal to the total energy (rest energy plus kinetic energy) divided by c^2. Because the relativistic mass is exactly proportional to the energy, relativistic mass and relativistic energy are nearly synonyms; the only difference between them is the units. If length and time are measured in natural units, the speed of light is equal to 1, and even this difference disappears. Then mass and energy have the same units and are always equal, so it is redundant to speak about relativistic mass, because it is just another name for the energy. This is why physicists usually reserve the useful short word “mass” to mean rest mass, or invariant mass, and not relativistic mass.

The relativistic mass of a moving object is larger than the relativistic mass of an object that is not moving, because a moving object has extra kinetic energy. The *rest mass* of an object is defined as the mass of an object when it is at rest, so that the rest mass is always the same, independent of the motion of the observer: it is the same in all inertial frames.

For things and systems made up of many parts, like an atomic nucleus, planet, or star, the relativistic mass is the sum of the relativistic masses (or energies) of the parts, because energies are additive in isolated systems. This is not true in open systems, however, if energy is subtracted. For example, if a system is *bound* by attractive forces, and the energy gained due to the forces of attraction in excess of the work done is removed from the system, then mass is lost with this removed energy. For example, the mass of an atomic nucleus is less than the total mass of the protons and neutrons that make it up, but this is only true after this energy from binding has been removed in the form of a gamma ray (which in this system, carries away the mass of the energy of binding). This mass decrease is also equivalent to the energy required to break up the nucleus into individual protons and neutrons (in this case, work and mass would need to be supplied). Similarly, the mass of the solar system is slightly less than the sum of the individual masses of the sun and planets.

For a system of particles going off in different directions, the invariant mass of the system is the analog of the rest mass, and is the same for all observers, even those in relative motion. It is defined as the total energy (divided by c^2) in the center of mass frame (where by definition, the system total momentum is zero). A simple example of an object with moving parts but zero total momentum is a container of gas. In this case, the mass of the container is given by its total energy (including the kinetic energy of the gas molecules), since the system total energy and invariant mass are the same in any reference frame where the momentum is zero, and such a reference frame is also the only frame in which the object can be weighed. In a similar way, the theory of special relativity posits that the thermal energy in all objects (including solids) contributes to their total masses and weights, even though this energy is present as the kinetic and potential energies of the atoms in the object, and it (in a similar way to the gas) is not seen in the rest masses of the atoms that make up the object.

In a similar manner, even photons (light quanta), if trapped in a container space (as a photon gas or thermal radiation), would contribute a mass associated with their energy to the container. Such an extra mass, in theory, could be weighed in the same way as any other type of rest mass. This is true in special relativity theory, even though individually photons have no rest mass. The property that trapped energy *in any form* adds weighable mass to systems that have no net momentum is one of the characteristic and notable consequences of relativity. It has no counterpart in classical Newtonian physics, in which radiation, light, heat, and kinetic energy never exhibit weighable mass under any circumstances.

Just as the relativistic mass of an isolated system is conserved through time, so also is its invariant mass.This property allows the conservation of all types of mass in systems, and also conservation of all types of mass in reactions where matter is destroyed (annihilated), leaving behind the energy that was associated with it (which is now in non-material form, rather than material form). Matter may appear and disappear in various reactions, but mass and energy are both unchanged in this process.

Applicability of the Strict Formula

Two different definitions of mass have been used in special relativity, and also two different definitions of energy. The simple equation $E = mc^2$ is not generally applicable to all these types of mass and energy, except in the special case that the total additive momentum is zero for the system under consideration. In such a case, which is always guaranteed when observing the system from either its center of mass frame or its center of momentum frame, $E = mc^2$ is always true for any type of mass and energy that are chosen. Thus, for example, in the center of mass frame, the total energy of an object or system is equal to its rest mass times $E = mc^2$, a useful equality. This is the relationship used for the container of gas in the previous example. It is *not* true in other reference frames where the center of mass is in motion. In these systems or for such an object, its total energy depends on both its rest (or invariant) mass, and its (total) momentum.

In inertial reference frames other than the rest frame or center of mass frame, the equation $E = mc^2$ remains true if the energy is the relativistic energy *and* the mass is the relativistic mass. It is also correct if the energy is the rest or invariant energy (also the minimum energy), *and* the mass is the rest mass, or the invariant mass. However, connection of the total or relativistic energy (E_r) with the rest or invariant mass (m_0) requires consideration of the system total momentum, in systems and reference frames where the total momentum (of magnitude *p*) has a non-zero value. The formula then required to connect the two different kinds of mass and energy, is the extended version of Einstein's equation, called the relativistic energy–momentum relation:

$$E_r^2 - |\vec{p}|^2 c^2 = m_0^2 c^4$$

$$E_r^2 - (pc)^2 = (m_0 c^2)^2$$

or

$$E_r = \sqrt{(m_0 c^2)^2 + (pc)^2}$$

Here the $(pc)^2$ term represents the square of the Euclidean norm (total vector length) of the various momentum vectors in the system, which reduces to the square of the simple momentum magnitude, if only a single particle is considered. This equation reduces to $E = mc^2$ when the momentum term is zero. For photons where $m_0 = 0$, the equation reduces to $E_r = pc$.

Meanings of the Strict Formula

Mass–energy equivalence states that any object has a certain energy, even when it is stationary. In Newtonian mechanics, a motionless body has no kinetic energy, and it may or may not have other amounts of internal stored energy, like chemical energy or thermal energy, in addition to any potential energy it may have from its position in a field of force. In Newtonian mechanics, all of these energies are much smaller than the mass of the object times the speed of light squared.

The mass–energy equivalence formula was displayed on Taipei 101 during the event of the World Year of Physics .

In relativity, all the energy that moves with an object (that is, all the energy present in the object's rest frame) contributes to the total mass of the body, which measures how much it resists acceleration. Each bit of potential and kinetic energy makes a proportional contribution to the mass. As noted above, even if a box of ideal mirrors "contains" light, then the individually massless photons still contribute to the total mass of the box, by the amount of their energy divided by c^2.

In relativity, removing energy is removing mass, and for an observer in the center of mass frame, the formula $m=\frac{E}{c^2}$ indicates how much mass is lost when energy is removed. In a nuclear reaction, the mass of the atoms that come out is less than the mass of the atoms that go in, and the difference in mass shows up as heat and light with the same relativistic mass as the difference (and also the same invariant mass in the center of mass frame of the system). In this case, the E in the formula is the energy released and removed, and the mass m is how much the mass decreases. In the same way, when any sort of energy is added to an isolated system, the increase in the mass is equal to the added energy divided by c^2. For example, when water is heated it gains about 1.11×10^{-17} kg of mass for every joule of heat added to the water.

An object moves with different speed in different frames, depending on the motion of the observer, so the kinetic energy in both Newtonian mechanics and relativity is *frame dependent*. This means that the amount of relativistic energy, and therefore the amount of relativistic mass, that an object is measured to have depends on the observer. The *rest mass* is defined as the mass that an object has when it is not moving (or when an inertial frame is chosen such that it is not moving). The term also applies to the invariant mass of systems when the system as a whole is not "moving" (has no net momentum). The rest and invariant masses are the smallest possible value of the mass of the object or system. They also are conserved quantities, so long as the system is isolated. Because of the way they are calculated, the effects of moving observers are subtracted, so these quantities do not change with the motion of the observer.

The rest mass is almost never additive: the rest mass of an object is not the sum of the rest masses of its parts. The rest mass of an object is the total energy of all the parts, including kinetic energy, as measured by an observer that sees the center of the mass of the object to be standing still. The rest mass adds up only if the parts are standing still and do not attract or repel, so that they do not have any extra kinetic or potential energy. The other possibility is that they have a positive kinetic energy and a negative potential energy that exactly cancels.

Binding Energy and the "Mass Defect"

Whenever any type of energy is removed from a system, the mass associated with the energy is also removed, and the system therefore loses mass. This mass defect in the system may be simply calculated as $\Delta m = \frac{\Delta E}{c^2}$, and this was the form of the equation historically first presented by Einstein in 1905. However, use of this formula in such circumstances has led to the false idea that mass has been "converted" to energy. This may be particularly the case when the energy (and mass) removed from the system is associated with the *binding energy* of the system. In such cases, the binding energy is observed as a "mass defect" or deficit in the new system.

The fact that the released energy is not easily weighed in many such cases, may cause its mass to be neglected as though it no longer existed. This circumstance has encouraged the false idea of conversion of *mass* to energy, rather than the correct idea that the binding energy of such systems is relatively large, and exhibits a measurable mass, which is removed when the binding energy is removed.

The difference between the rest mass of a bound system and of the unbound parts is the binding energy of the system, if this energy has been removed after binding. For example, a water molecule weighs a little less than two free hydrogen atoms and an oxygen atom. The minuscule mass difference is the energy needed to split the molecule into three individual atoms (divided by c^2), which was given off as heat when the molecule formed (this heat had mass). Likewise, a stick of dynamite in theory weighs a little bit more than the fragments after the explosion, but this is true only so long as the fragments are cooled and the heat removed. In this case the mass difference is the energy/heat that is released when the dynamite explodes, and when this heat escapes, the mass associated with it escapes, only to be deposited in the surroundings, which absorb the heat (so that total mass is conserved).

Such a change in mass may only happen when the system is open, and the energy and mass escapes. Thus, if a stick of dynamite is blown up in a hermetically sealed chamber, the mass of the chamber and fragments, the heat, sound, and light would still be equal to the original mass of the chamber and dynamite. If sitting on a scale, the weight and mass would not change. This would in theory also happen even with a nuclear bomb, if it could be kept in an ideal box of infinite strength, which did not rupture or pass radiation. Thus, a 21.5 kiloton (9×10^{13} joule) nuclear bomb produces about one gram of heat and electromagnetic radiation, but the mass of this energy would not be detectable in an exploded bomb in an ideal box sitting on a scale; instead, the contents of the box would be heated to millions of degrees without changing total mass and weight. If then, however, a transparent window (passing only electromagnetic radiation) were opened in such an ideal box after the explosion, and a beam of X-rays and other lower-energy light allowed to escape the box, it would eventually be found to weigh one gram less than it had before the explosion. This weight loss and mass loss would happen as the box was cooled by this process, to room temperature. However, any surrounding mass that absorbed the X-rays (and other "heat") would *gain* this gram of mass from the resulting heating, so the mass "loss" would represent merely its relocation. Thus, no mass (or, in the case of a nuclear bomb, no matter) would be "converted" to energy in such a process. Mass and energy, as always, would both be separately conserved.

Massless Particles

Massless particles have zero rest mass. Their relativistic mass is simply their relativistic energy, divided

By c^2, or $m_{rel} = \frac{E}{c^2}$. The energy for photons is $E = hf$, where h is Planck's constant and f is the photon frequency. This frequency and thus the relativistic energy are frame-dependent.

If an observer runs away from a photon in the direction the photon travels from a source, and it catches up with the observer—when the photon catches up, the observer sees it as having less energy than it had at the source. The faster the observer is traveling with regard to the source when the photon catches up, the less energy the photon has. As an observer approaches the speed of light with regard to the source, the photon looks redder and redder, by relativistic Doppler effect (the Doppler shift is the relativistic formula), and the energy of a very long-wavelength photon approaches zero. This is because the photon is *massless*—the rest mass of a photon is zero.

Massless Particles Contribute Rest Mass and Invariant Mass to Systems

Two photons moving in different directions cannot both be made to have arbitrarily small total energy by changing frames, or by moving toward or away from them. The reason is that in a two-photon system, the energy of one photon is decreased by chasing after it, but the energy of the other increases with the same shift in observer motion. Two photons not moving in the same direction comprise an inertial frame where the combined energy is smallest, but not zero. This is called the center of mass frame or the center of momentum frame; these terms are almost synonyms (the center of mass frame is the special case of a center of momentum frame where the center of mass is put at the origin). The most that chasing a pair of photons can accomplish to decrease their energy is to put the observer in a frame where the photons have equal energy and are moving directly away from each other. In this frame, the observer is now moving in the same direction and speed as the center of mass of the two photons. The total momentum of the photons is now zero, since their momenta are equal and opposite. In this frame the two photons, as a system, have a mass equal to their total energy divided by c^2. This mass is called the invariant mass of the pair of photons together. It is the smallest mass and energy the system may be seen to have, by any observer. It is only the invariant mass of a two-photon system that can be used to make a single particle with the same rest mass.

If the photons are formed by the collision of a particle and an antiparticle, the invariant mass is the same as the total energy of the particle and antiparticle (their rest energy plus the kinetic energy), in the center of mass frame, where they automatically move in equal and opposite directions (since they have equal momentum in this frame). If the photons are formed by the disintegration of a *single* particle with a well-defined rest mass, like the neutral pion, the invariant mass of the photons is equal to rest mass of the pion. In this case, the center of mass frame for the pion is just the frame where the pion is at rest, and the center of mass does not change after it disintegrates into two photons. After the two photons are formed, their center of mass is still moving the same way the pion did, and their total energy in this frame adds up to the mass energy of the pion. Thus, by calculating the invariant mass of pairs of photons in a particle detector, pairs can be identified that were probably produced by pion disintegration.

A similar calculation illustrates that the invariant mass of systems is conserved, even when massive particles (particles with rest mass) within the system are converted to massless particles (such as photons). In such cases, the photons contribute invariant mass to the system, even though they individually have no invariant mass or rest mass. Thus, an electron and positron (each of which has rest mass) may undergo annihilation with each other to produce two photons, each of which is massless (has no rest mass). However, in such circumstances, no system mass is lost. Instead, the system of both photons moving away from each other has an invariant mass, which acts like a rest mass for any system in which the photons are trapped, or that can be weighed. Thus, not only the quantity of relativistic mass, but also the quantity of invariant mass does not change in transformations between "matter" (electrons and positrons) and energy (photons).

Relation to Gravity

In physics, there are two distinct concepts of mass: the gravitational mass and the inertial mass. The gravitational mass is the quantity that determines the strength of the gravitational field generated by an object, as well as the gravitational force acting on the object when it is immersed in a gravitational field produced by other bodies. The inertial mass, on the other hand, quantifies how much an object accelerates if a given force is applied to it. The mass–energy equivalence in special relativity refers to the inertial mass. However, already in the context of Newton gravity, the Weak Equivalence Principle is postulated: the gravitational and the inertial mass of every object are the same. Thus, the mass–energy equivalence, combined with the weak equivalence principle, results in the prediction that all forms of energy contribute to the gravitational field generated by an object. This observation is one of the pillars of the general theory of relativity.

The above prediction, that all forms of energy interact gravitationally, has been subject to experimental tests. The first observation testing this prediction was made in 1919. During a solar eclipse, Arthur Eddington observed that the light from stars passing close to the Sun was bent. The effect is due to the gravitational attraction of light by the Sun. The observation confirmed that the energy carried by light indeed is equivalent to a gravitational mass. Another seminal experiment, the Pound–Rebka experiment, was performed in 1960. In this test a beam of light was emitted from the top of a tower and detected at the bottom. The frequency of the light detected was higher than the light emitted. This result confirms that the energy of photons increases when they fall in the gravitational field of the Earth. The energy, and therefore the gravitational mass, of photons is proportional to their frequency as stated by the Planck's relation.

Application to Nuclear Physics

Task Force One, the world's first nuclear-powered task force. *Enterprise*, *Long Beach* and *Bainbridge* in formation in the Mediterranean, 18 June 1964. *Enterprise* crew members are spelling out Einstein's mass–energy equivalence formula $E = mc^2$ on the flight deck.

Max Planck pointed out that the mass–energy equivalence formula implied that bound systems would have a mass less than the sum of their constituents, once the binding energy had been allowed to escape. However, Planck was thinking about chemical reactions, where the binding energy is too small to measure. Einstein suggested that radioactive materials such as radium would provide a test of the theory, but even though a large amount of energy is released per atom in radium, due to the half-life of the substance (1602 years), only a small fraction of radium atoms decay over an experimentally measurable period of time.

Once the nucleus was discovered, experimenters realized that the very high binding energies of the atomic nuclei should allow calculation of their binding energies, simply from mass differences. But it was not until the discovery of the neutron in 1932, and the measurement of the neutron mass, that this calculation could actually be performed. A little while later, the Cockcroft–Walton accelerator produced the first transmutation reaction (${}^{7}_{3}Li + {}^{1}_{1}p \rightarrow 2\,{}^{4}_{2}He$), verifying Einstein's formula to an accuracy of ±0.5%. In 2005, Rainville et al. published a direct test of the energy-equivalence of mass lost in the binding energy of a neutron to atoms of particular isotopes of silicon and sulfur, by comparing the mass lost to the energy of the emitted gamma ray associated with the neutron capture. The binding mass-loss agreed with the gamma ray energy to a precision of ±0.00004%, the most accurate test of $E = mc^2$ to date.

The mass–energy equivalence formula was used in the understanding of nuclear fission reactions, and implies the great amount of energy that can be released by a nuclear fission chain reaction, used in both nuclear weapons and nuclear power. By measuring the mass of different atomic nuclei and subtracting from that number the total mass of the protons and neutrons as they would weigh separately, one gets the exact binding energy available in an atomic nucleus. This is used to calculate the energy released in any nuclear reaction, as the difference in the total mass of the nuclei that enter and exit the reaction.

Efficiency

Although mass cannot be converted to energy, in some reactions matter particles (which contain a form of rest energy) can be destroyed and the energy released can be converted to other types of energy that are more usable and obvious as forms of energy—such as light and energy of motion (heat, etc.). However, the total amount of energy and mass does not change in such a transformation. Even when particles are not destroyed, a certain fraction of the ill-defined "matter" in ordinary objects can be destroyed, and its associated energy liberated and made available as the more dramatic energies of light and heat, even though no identifiable real particles are destroyed, and even though (again) the total energy is unchanged (as also the total mass). Such conversions between types of energy (resting to active energy) happen in nuclear weapons, in which the protons and neutrons in atomic nuclei lose a small fraction of their average mass, but this mass loss is not due to the destruction of any protons or neutrons (or even, in general, lighter particles like electrons). Also the mass is not destroyed, but simply removed from the system in the form of heat and light from the reaction.

In nuclear reactions, typically only a small fraction of the total mass–energy of the bomb converts into the mass–energy of heat, light, radiation, and motion—which are "active" forms that can be used. When an atom fissions, it loses only about 0.1% of its mass (which escapes from the system and does not disappear), and additionally, in a bomb or reactor not all the atoms can fission. In a modern fission-based atomic bomb, the efficiency is only about 40%, so only 40% of the fissionable atoms actually fission, and only about 0.03% of the fissile core mass appears as energy in the end. In nuclear fusion, more of the mass is released as usable energy, roughly 0.3%. But in a fusion bomb, the bomb mass is partly casing and non-reacting components, so that in practicality, again (coincidentally) no more than about 0.03% of the total mass of the entire weapon is released as usable energy (which, again, retains the "missing" mass).

In theory, it should be possible to destroy matter and convert all of the rest-energy associated with matter into heat and light (which would of course have the same mass), but none of the theoretically known methods are practical. One way to convert all the energy within matter into usable energy is to annihilate matter with antimatter. But antimatter is rare in our universe, and must be made first. Due to inefficient mechanisms of production, making antimatter always requires far more usable energy than would be released when it was annihilated.

Since most of the mass of ordinary objects resides in protons and neutrons, converting all the energy of ordinary matter into more useful energy requires that the protons and neutrons be converted to lighter particles, or particles with no rest-mass at all. In the Standard Model of particle physics, the number of protons plus neutrons is nearly exactly conserved. Still, Gerard 't Hooft showed that there is a process that converts protons and neutrons to antielectrons and neutrinos. This is the weak SU(2) instanton proposed by Belavin Polyakov Schwarz and Tyupkin. This process, can in principle destroy matter and convert all the energy of matter into neutrinos and usable energy, but it is normally extraordinarily slow. Later it became clear that this process happens at a fast rate at very high temperatures, since then, instanton-like configurations are copiously produced from thermal fluctuations. The temperature required is so high that it would only have been reached shortly after the big bang.

Many extensions of the standard model contain magnetic monopoles, and in some models of grand unification, these monopoles catalyze proton decay, a process known as the Callan-Rubakov effect. This process would be an efficient mass–energy conversion at ordinary temperatures, but it requires making monopoles and anti-monopoles first. The energy required to produce monopoles is believed to be enormous, but magnetic charge is conserved, so that the lightest monopole is stable. All these properties are deduced in theoretical models—magnetic monopoles have never been observed, nor have they been produced in any experiment so far.

A third known method of total matter–energy "conversion" (which again in practice only means conversion of one type of energy into a different type of energy), is using gravity, specifically black holes. Stephen Hawking theorized that black holes radiate thermally with no regard to how they are formed. So, it is theoretically possible to throw matter into a black hole and use the emitted heat to generate power. According to the theory of Hawking radiation, however, the black hole used radiates at a higher rate the smaller it is, producing usable powers at only small black hole masses, where usable may for example be something greater than the local background radiation. It is also worth noting that the ambient irradiated power would change with the mass of the black hole, increasing as the mass of the black hole decreases, or decreasing as the mass increases, at a rate where power is proportional to the inverse square of the mass. In a "practical" scenario, mass and energy could be dumped into the black hole to regulate this growth, or keep its size, and thus power output, near constant. This could result from the fact that mass and energy are lost from the hole with its thermal radiation.

Nuclear Binding Energy

The energy which holds the nucleons in the nucleus and makes it stabilize is called as nuclear bonding energy. Nucleus consists of protons and neutrons hence the mass of nucleus must be sum of mass of neutrons and protons. But the mass of a nucleus is always less than the sum of the individual masses of the nucleons. The nuclear binding energy can be defined as the energy needs to hold nucleons in a nucleus. The difference in mass of nucleus from the theoretical value is measure of nuclear binding energy. The Einstein relationship can be used to determine the nuclear binding energy that is as given below;

Nuclear binding energy= Δmc^2 .

Here Δm is mass defect. If we know the magnitude of mass defect for certain particle, we can calculate the nuclear bonding energy for that nucleus such as the value of Δm for alpha particles is 0.0304 u, hence the nuclear binding energy for same would be 28.3 MeV. Remember the nuclear bonding energy is different from the electron bonding energy which is the binding energy of an electron in an atom. The magnitude of nuclear binding energies is much greater than the electron binding energies of atoms. So we can say that the energy needed to dissociate the nucleus into its component nucleons is known as nuclear binding energy. It is usually expressed in terms of kJ/mole of nuclei or MeV's/nucleon.

Causes of Binding Energy

We know that the nucleus of an atom is formed by the combination of protons and neutrons which are also known as nucleons. The atoms which have almost same number of protons and neutrons are more stable compare than those with odd numbers of these subatomic particles. In other words we can say that nuclides which have number of protons and neutrons equal to 2, 8, 20, 50,

82 or 126; also known as "magic numbers" are especially stable. The n/p ratio of atoms is useful to explain the nuclear stability but it is not a quantitative measure of the nuclear stability. Since a nucleus is composed of protons and neutrons therefore the sum of masses of neutrons and protons must be the mass of nucleus. But the mass of a nucleus is little less than the mass of nucleons. This different is called as mass defects.

Δm = Mass defect = Calculated mass - Actual mass

This difference in mass of nucleus is due to some energy change in the nucleus which provides stability to the nucleus and called as nuclear binding energy.

Nuclear Binding Energy Formula

Mass defect is difference between the mass of a nucleus and the sum of the masses of the nucleons. Hence for the calculation of mass defect we need the actual mass of the nucleus and also the total number of protons and neutrons with masses of them. First add up the masses of each proton and of each neutron then subtract the actual mass of the nucleus. Now we have to convert this mass defect into energy units and for that first we have to convert the mass defect into kilograms and then convert the mass defect into its energy equivalent with the help of Einstein's equation. We can also convert the nuclear binding energy in kJ/mol of nuclei or as MeV's/nucleon.

Nuclear Binding Energy Equation

The Einstein relation of mass and energy explains that matter and energy are equivalent. Same relation can be used to determine the nuclear binding energy. The mass defect is the amount of matter which would be converted into energy during the formation of a nucleus from its component nucleons. The energy value is known as nuclear binding energy and the nuclear bonding energy equation can be written as;

$$BE = (\Delta m)c^2$$

Nuclear Binding Energy Curve

The nuclear binding energy curve represents the amount of binding energy per nucleon. Here nucleon could be either a neutron or a proton. We know that the nucleon number is the sum of the number of neutrons and protons in a nucleus. Hence it is also equal to the mass number of an atom. In this plot, higher curve indicates the stability of nucleus. We can notice the peak at A = 60 it means the nuclei with A = 60 and around this number are very stable in nature. This peak is also called as *iron peak nuclei.*

Fission and fusion reactions can be explained with the help of nuclear bonding curve. The plot suggests that the heaviest nuclei are less stable. The mass of these unstable nuclei is more than 60. It indicates that the energy can be released during the fission of heavy nuclei. Similarly we can predict the fusion reactions with the help of these curves. The nuclear binding energy curves also explained that light weight elements like Helium or Hydrogen are less stable than heavier elements up to mass number 60. Therefore fusion of these nuclei can release energy.

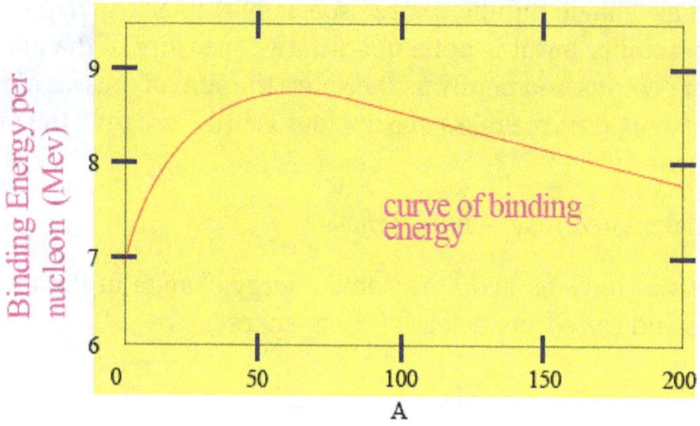

Total Binding Energy

In many respects the nucleus behaves like a nonpolar liquid drop. To examine the stability of such a nuclear system, it is useful to examine the Total Binding Energy, which is the mass converted into energy when a nucleus is formed from its constituent nucleons,

$$Z^1H + N^1n \rightarrow {}^A_Z X_N + \frac{TBE}{c^2} u.$$

For which the mass balance equation is

$$TBE = \left[ZM_H + N\,M_N - M\left({}^A_Z X\right)\right] c^2 \text{ MeV}.$$

Substituting Eq. $M = A + \Delta / C^2$ for M,

$$TBE = \left[Z\Delta_H + N\,\Delta_N - \Delta\left({}^A_Z X\right)\right] c^2 \text{ MeV}.$$

The total binding energy is analogous to the heat of condensation of a liquid and the reverse equation corresponds to the heat of vaporization. The TBE for nuclei ranges from 2.2 MeV for the simplest nucleus, 2H, to ~2000 MeV for uranium nuclei.

A more instructive quantity is the average binding energy per nucleon, <BE>, analogous to a molar heat of condensation (other than a factor of 6.023 x 1023, to account for the number of particles in a mole instead of a single particle).

$$< BE > = TBE / A$$

The average binding energy is the amount of energy required to remove a nucleon from the nucleus, assuming all nucleons are equally bound (true for an ideal nonpolar liquid and to a fair approximation for nuclei).

Exercise: Calculate <BE> for ^{12}C.

$$6\,{}^{1}H + 6\,{}^{1}n \rightarrow {}^{12}C + TBE/c^2$$

$$TBE = 6\Delta_H + 6\Delta_n - \Delta({}^{12}C)$$

$$= 6(7.289) + 6(8.071) - 0 = 92.160\ MeV$$

$$\langle BE \rangle = 92.160/12 = 7.680\ MeV$$

Nuclear Energetics: Particle Binding Energies

The particle binding energy is the energy required to remove a particle from the nucleus. As the binding energy of a particle (proton, neutron, etc.) decreases, the nucleus becomes increasingly unstable toward spontaneous emission of that particle. This definition is comparable to the case for electron binding energies in an atom, binding energy, Bi, is defined as separation energy, S_i. Here we summarize the three most relevant cases:

- Proton binding energy, B_p – the energy required to remove a proton from a nucleus,

  EQUATION: ${}^{A}_{Z}X + B_\rho c^2 \rightarrow {}^{A-1}_{Z-1}Y + {}^{1}_{1}H$

  CALCULATION: $B_p = \Delta(Y) + \Delta(H) - \Delta(X)$

- Neutron binding energy, B_n – the energy required to remove a neutron from a nucleus,

  EQUATION: ${}^{A}_{Z}X + B_n c^2 \rightarrow {}^{A-1}_{Z}X + {}^{1}_{0}n$

  CALCULATION: $B_n = \Delta({}^{A-1}X) + \Delta n - \Delta({}^{A}X)$

- Alpha-particle (${}^{4}He$) binding energy, B_α.. the energy required to remove a ${}^{4}He$ from a nucleus,

  EQUATION: ${}^{A}_{Z}X + B_\alpha c^2 \rightarrow {}^{A-4}_{Z-2}Y + {}^{4}_{2}He$

  CALCULATION: $B_\alpha = (\Delta Y) + \Delta(\alpha) - \Delta(X)$

In principle, one could calculate the binding energy for any particle: e.g. ${}^{2}H$, ${}^{14}C$, etc.

Example: Calculate the neutron binding energy for 12Be

$${}^{12}Be + Bn\,c^2 \rightarrow {}^{11}Be + {}^{1}n$$

$$B_n = \Delta({}^{11}Be) + \Delta({}^{1}n) - \Delta({}^{12}Be)$$

$$B_n = 20.174 + 8.071 - 25.007 = 3.168\,MeV$$

Nuclear Fusion

The binding energy of helium is the energy source of the Sun and of most stars. The sun is composed of 74 percent hydrogen (measured by mass), an element whose nucleus is a single proton. Energy is released in the sun when 4 protons combine into a helium nucleus, a process in which two of them are also converted to neutrons.

The conversion of protons to neutrons is the result of another nuclear force, known as the weak (nuclear) force. The weak force, like the strong force, has a short range, but is much weaker than the strong force. The weak force tries to make the number of neutrons and protons into the most energetically stable configuration. For nuclei containing less than 40 particles, these numbers are usually about equal. Protons and neutrons are closely related and are collectively known as nucleons. As the number of particles increases toward a maximum of about 209, the number of neutrons to maintain stability begins to outstrip the number of protons, until the ratio of neutrons to protons is about three to two.

The protons of hydrogen combine to helium only if they have enough velocity to overcome each other's mutual repulsion sufficiently to get within range of the strong nuclear attraction. This means that fusion only occurs within a very hot gas. Hydrogen hot enough for combining to helium requires an enormous pressure to keep it confined, but suitable conditions exist in the central regions of the Sun, where such pressure is provided by the enormous weight of the layers above the core, pressed inwards by the Sun's strong gravity. The process of combining protons to form helium is an example of nuclear fusion.

The earth's oceans contain a large amount of hydrogen that could theoretically be used for fusion, and helium byproduct of fusion does not harm the environment, so some consider nuclear fusion a good alternative to supply humanity's energy needs. Experiments to generate electricity from fusion have so far only partially succeeded. Sufficiently hot hydrogen must be ionized and confined. One technique is to use very strong magnetic fields, because charged particles (like those trapped in the Earth's radiation belt) are guided by magnetic field lines. Fusion experiments also rely on heavy hydrogen, which fuses more easily, and gas densities can be moderate. But even with these techniques far more net energy is consumed by the fusion experiments than is yielded by the process.

Binding Energy Maximum and Ways to Approach it by Decay

In the main isotopes of light nuclei, such as carbon, nitrogen and oxygen, the most stable combination of neutrons and of protons are when the numbers are equal (this continues to element 20, calcium). However, in heavier nuclei, the disruptive energy of protons increases, since they are confined to a tiny volume and repel each other. The energy of the strong force holding the nucleus together also increases, but at a slower rate, as if inside the nucleus, only nucleons close to each other are tightly bound, not ones more widely separated.

The net binding energy of a nucleus is that of the nuclear attraction, minus the disruptive energy of the electric force. As nuclei get heavier than helium, their net binding energy per nucleon (deduced from the difference in mass between the nucleus and the sum of masses of component nucleons) grows more and more slowly, reaching its peak at iron. As nucleons are added, the total nuclear binding energy always increases—but the total disruptive energy of electric forces (positive protons repelling other protons) also increases, and past iron, the second increase outweighs the first. Iron-56 (^{56}Fe) is the most efficiently bound nucleus meaning that it has the least average mass per nucleon. However, nickel-62 is the most tightly bound nucleus in terms of energy of binding per nucleon . (Nickel-62's higher energy of binding does not translate to a larger mean mass loss than Fe-56, because Ni-62 has a slightly higher ratio of neutrons/protons than does iron-56, and the presence of the heavier neutrons increases nickel-62's average mass per nucleon).

To reduce the disruptive energy, the weak interaction allows the number of neutrons to exceed that of protons—for instance, the main isotope of iron has 26 protons and 30 neutrons. Isotopes also exist where the number of neutrons differs from the most stable number for that number of nucleons. If the ratio of protons to neutrons is too far from stability, nucleons may spontaneously change from proton to neutron, or neutron to proton.

The two methods for this conversion are mediated by the weak force, and involve types of beta decay. In the simplest beta decay, neutrons are converted to protons by emitting a negative electron and an antineutrino. This is always possible outside a nucleus because neutrons are more massive than protons by an equivalent of about 2.5 electrons. In the opposite process, which only happens within a nucleus, and not to free particles, a proton may become a neutron by ejecting a positron. This is permitted if enough energy is available between parent and daughter nuclides to do this (the required energy difference is equal to 1.022 MeV, which is the mass of 2 electrons). If the mass difference between parent and daughter is less than this, a proton-rich nucleus may still convert protons to neutrons by the process of electron capture, in which a proton simply electron captures one of the atom's K orbital electrons, emits a neutrino, and becomes a neutron.

Among the heaviest nuclei, starting with tellurium nuclei (element 52) containing 106 or more nucleons, electric forces may be so destabilizing that entire chunks of the nucleus may be ejected, usually as alpha particles, which consist of two protons and two neutrons (alpha particles are fast helium nuclei). (Beryllium-8 also decays, very quickly, into two alpha particles.) Alpha particles are extremely stable. This type of decay becomes more and more probable as elements rise in atomic weight past 106.

The curve of binding energy is a graph that plots the binding energy per nucleon against atomic mass. This curve has its main peak at iron and nickel and then slowly decreases again, and also a narrow isolated peak at helium, which as noted is very stable. The heaviest nuclei in nature, uranium ^{238}U, are unstable, but having a half-life of 4.5 billion years, close to the age of the Earth, they are still relatively abundant; they (and other nuclei heavier than helium) have formed in stellar evolution events like supernova explosions preceding the formation of the solar system. The most common isotope of thorium, ^{232}Th, also undergoes alpha particle emission, and its half-life (time over which half a number of atoms decays) is even longer, by several times. In each of these, radioactive decay produces daughter isotopes that are also unstable, starting a chain of decays that ends in some stable isotope of lead.

Binding Energy for Atoms

The binding energy of an atom (including its electrons) is not the same as the binding energy of the atom's nucleus. The measured mass deficits of isotopes are always listed as mass deficits of the neutral atoms of that isotope, and mostly in MeV. As a consequence, the listed mass deficits are not a measure for the stability or binding energy of isolated nuclei, but for the whole atoms. This has very practical reasons, because it is very hard to totally ionize heavy elements, i.e. strip them of all of their electrons.

This practice is useful for other reasons, too: stripping all the electrons from a heavy unstable nucleus (thus producing a bare nucleus) changes the lifetime of the nucleus, or the nucleus of a stable neutral atom can likewise become unstable after stripping, indicating that the nucleus cannot be treated independently. Examples of this have been shown in bound-state β decay experiments performed at the GSI) heavy ion accelerator. This is also evident from phenomena like electron capture. Theoretically, in orbital models of heavy atoms, the electron orbits partially inside the nucleus (it does not *orbit* in a strict sense, but has a non-vanishing probability of being located inside the nucleus).

A nuclear decay happens to the nucleus, meaning that properties ascribed to the nucleus change in the event. In the field of physics the concept of "mass deficit" as a measure for "binding energy" means "mass deficit of the neutral atom" (not just the nucleus) and is a measure for stability of the whole atom.

Binding Energy and Nuclide Masses

The fact that the maximum binding energy is found in medium-sized nuclei is a consequence of the trade-off in the effects of two opposing forces that have different range characteristics. The attractive nuclear force (strong nuclear force), which binds protons and neutrons equally to each other, has a limited range due to a rapid exponential decrease in this force with distance. However, the repelling electromagnetic force, which acts between protons to force nuclei apart, falls off with distance much more slowly (as the inverse square of distance). For nuclei larger than about four nucleons in diameter, the additional repelling force of additional protons more than offsets any binding energy that results between further added nucleons as a result of additional strong force interactions. Such nuclei become increasingly less tightly bound as their size increases, though most of them are still stable. Finally, nuclei containing more than 209 nucleons (larger than about 6 nucleons in diameter) are all too large to be stable, and are subject to spontaneous decay to smaller nuclei.

Nuclear fusion produces energy by combining the very lightest elements into more tightly bound elements (such as hydrogen into helium), and nuclear fission produces energy by splitting the heaviest elements (such as uranium and plutonium) into more tightly bound elements (such as barium and krypton). Both processes produce energy, because middle-sized nuclei are the most tightly bound of all.

As seen above in the example of deuterium, nuclear binding energies are large enough that they may be easily measured as fractional mass deficits, according to the equivalence of mass and energy. The atomic binding energy is simply the amount of energy (and mass) released, when a collection of free nucleons are joined together to form a nucleus.

Nuclear binding energy can be computed from the difference in mass of a nucleus, and the sum of the masses of the number of free neutrons and protons that make up the nucleus. Once this mass difference, called the mass defect or mass deficiency, is known, Einstein's mass-energy equivalence formula $E = mc^2$ can be used to compute the binding energy of any nucleus. Early nuclear physicists used to refer to computing this value as a "packing fraction" calculation.

For example, the atomic mass unit (1 u) is defined as 1/12 of the mass of a ^{12}C atom—but the atomic mass of a ^{1}H atom (which is a proton plus electron) is 1.007825 *u*, so each nucleon in ^{12}C has lost, on average, about 0.8% of its mass in the form of binding energy.

Semi-empirical Mass Formula

$$E_b(\text{MeV}) = a_V A - a_s A^{\frac{2}{3}} - a_c \frac{Z^2}{A^{\frac{1}{3}}} - a_c \frac{(A-2Z)^2}{A} \pm \delta(A,Z)$$

$$+\delta_0 \ for\ Z,N\ even$$
$$\delta(A,Z) = 0$$
$$-\delta_0 \ for\ Z,N\ odd$$

This formula is called the Weizsaecker Formula (or the semi-empirical mass formula). The physical meaning of this equation can be discussed term by term.

Volume Term

Volume term – $a_V.A$. The first two terms describe a spherical liquid drop of an incompressible fluid with a contribution from the volume scaling with A and from the surface, scaling with $A^{2/3}$. The first positive term $a_V.A$ is known as the volume term and it is caused by the attracting strong forces between the nucleons. The strong force has a very limited range and a given nucleon may only interact with its direct neighbours. Therefore this term is proportional to A, instead of A^2. The coefficient a_V is usually about ~ 16 MeV.

Surface Term

Surface term – $a_{sf}.A^{2/3}$. The surface term is also based on the strong force, it is, in fact, a correction to the volume term. The point is that particles at the surface of the nucleus are not completely surrounded by other particles. In the volume term, it is suggested that each nucleon interacts with a constant number of nucleons, independent of A. This assumption is very nearly true for nucleons deep within the nucleus, but causes an overestimation of the binding energy on the surface. By analogy with a liquid drop this effect is indicated as the surface tension effect. If the volume of the nucleus is proportional to A, then the geometrical radius should be proportional to $A^{1/3}$ and therefore the surface term must be proportional to the surface area i.e. proportional to $A^{2/3}$.

Coulomb Term

Coulomb term – $a_C.Z^2.A^{-1/3}$. This term describes the Coulomb repulsion between the uniformly distributed protons and is proportional to the number of proton pairs Z^2/R, whereby R is proportional to $A^{1/3}$. This effect lowers the binding energy because of the repulsion between charges of equal sign.

Asymmetry Term

Asymmetry term – $a_A.(A-2Z)^2/A$. This term cannot be described as 'classically' as the first three. This effect is not based on any of the fundamental forces, this effect is based only on the Pauli exclusion principle (no two fermions can occupy exactly the same quantum state in an atom). The heavier nuclei contain more neutrons than protons. These extra neutrons are necessary for stability of the heavier nuclei. They provide (via the attractive forces between the neutrons and protons) some compensation for the repulsion between the protons. On the other hand, if there are significantly more neutrons than protons in a nucleus, some of the neutrons will be higher in energy level in the nucleus. This is the basis for a correction factor, the so-called symmetry term.

Pairing Term

Pairing term – δ(A,Z). The last term is the pairing term δ(A,Z). This term captures the effect of spin-coupling. Nuclei with an even number of protons and an even number of neutrons are (due to Pauli exclusion principle) very stable thanks to the occurrence of 'paired spin'. On the other hand, nuclei with an odd number of protons and neutrons are mostly unstable.

Table of Calculated Binding Energies

	$^{40}_{20}Ca$	$^{107}_{47}Ag$	$^{238}_{92}U$
volume term	630	1686	3751
surface term	-208	-401	-684
coulomb term	-83	-331	-971
symmetry term	0	-37	-290
pairing term	+2	0	+0.6
calculated E_B	341	917	1806
measured E_B	342	915	1802
measured E_B/A	8.6	8.6	7.6

Table of binding energies fo some nuclides. Calculated according to the semi-empirical mass formula.

With the aid of the Weizsaecker formula the binding energy can be calculated very well for nearly all isotopes. This formula provides a good fit for heavier nuclei. For light nuclei, especially for ^{4}He, it provides a poor fit. The main reason is the formula does not consider the internal shell structure of the nucleus.

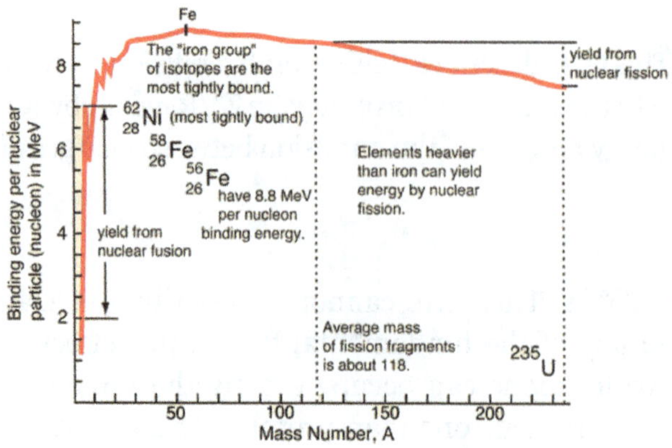

Nuclear binding energy curve.

In order to calculate the binding energy, the coefficients a_V, a_S, a_C, a_A and a_P must be known. The coefficients have units of megaelectronvolts (MeV) and are calculated by fitting to experimentally measured masses of nuclei. They usually vary depending on the fitting methodology. According to ROHLF, J. W., Modern Physics from α to Zo, the coefficients in the equation are following:

$$E_b(\text{MeV}) = 15.76A - 17.81A^{2/3} - 0.711\frac{Z^2}{A^{1/3}} - 23.7\frac{(N-Z)^2}{A} \pm 34A^{-3/4}$$

Using the Weizsaecker formula, also the mass of an atomic nucleus can be derived and is given by:$m = Z.m_p + N.m_n - E_b/c^2$

where m_p and m_n are the rest mass of a proton and a neutron, respectively, and E_b is the nuclear binding energy of the nucleus.

From the nuclear binding energy curve and from the table it can be seen that, in the case of splitting a ^{235}Unucleus into two parts, the binding energy of the fragments (A ≈ 120) together is larger than that of the original ^{235}U nucleus.

According to the Weizsaecker formula, the total energy released for such reaction will be approximately 235 x (8.5 – 7.6) ≈ 200 MeV.

Q-value

The Q value of a nuclear reaction is the difference between the sum of the masses of the initial reactants and the sum of the masses of the final products, in energy units (usually in MeV). This is also the corresponding difference of the binding energies of the nuclei (not per nucleon), since nucleon number is conserved in a reaction. The masses may be provided in a table of mass excesses $(\Delta M(A, Z))$, which is the value of $M(A, Z) - Am_u$ (usually in MeV), but relative to the corresponding number for the isotope ^{12}C. $(m_u \equiv m_{amu}.)$ Hence, the "mass excess" of ^{12}C is defined to be zero. For instance, for the reaction $^{12}C(\alpha, \gamma)\ ^{16}O$, the Q value is:

$$Q = 931.478 \text{ MeV } M(^{12}C) + M(^{4}He) - M(^{16}O)$$
$$= 7.1613 \text{ MeV}$$

where the mass excess of ^{16}O is negative (oxygen is more bound than carbon). For the triple-α process, one finds Q/A = 0.606 MeV/nucleon, for hydrogen burning it is 26.73 MeV, or ~6.7 MeV per nucleon. Note that 1 MeV/nucleon is equivalent to 0.965 × 1018 erg g^{-1}, a very useful conversion factor. Note also that for hydrogen burning the efficiency of conversion of mass into energy is ~6.7/931≈0.007, less than but near 1%. This is the core fact of fusion as the source of energy for stars.

Consider a typical reaction, in which the projectile a and the target A gives place to two products,

B and b. This can also be expressed in the notation that we used so far, a + A → B + b, or even in a more compact notation, A(a, b)B.

The Q-value of this reaction is given by:

$$Q = [m_a + m_A - (m_b + m_B)]c^2$$

which is the same as the excess kinetic energy of the final products:

$$\begin{aligned} Q &= T_{final} - T_{initial} \\ &= T_b + T_B - (T_a + T_A) \end{aligned}$$

For reactions in which there is an increase in the kinetic energy of the products Q is positive. The positive Q reactions are said to be exothermic (or exergic). There is a net release of energy, since the kinetic energy of the final state is greater than the kinetic energy of the initial state.

For reactions in which there is a decrease in the kinetic energy of the products Q is negative. The negative Q reactions are said to be endothermic (or endoergic) and they require a net energy input.

Q-value of Exothermic Reactions

Example: Exothermic Reaction - DT fusion

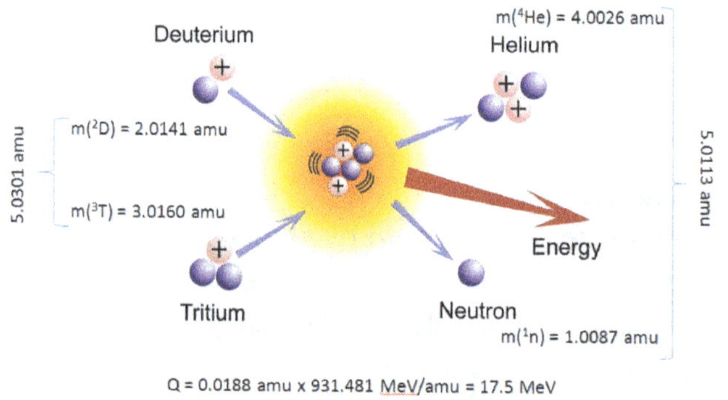

Q-value of DT fusion reaction

The DT fusion reaction of deuterium and tritium is particularly interesting because of its potential of providing energy for the future. Calculate the reaction Q-value.

3T (d, n) 4He

The atom masses of the reactants and products are:

$$\begin{aligned} m(^3T) &= 3.0160\ amu \\ m(^2D) &= 2.0141\ amu \\ m(^1n) &= 1.0087\ amu \\ m(^4He) &= 4.0026\ amu \end{aligned}$$

Using the mass-energy equivalence, we get the Q-value of this reaction as:

Q = {(3.0160+2.0141) [amu] – (1.0087+4.0026) [amu]} x 931.481 [MeV/amu]

= 0.0188 x 931.481 = 17.5 MeV

Example: Exothermic Reaction - Tritium in Reactors

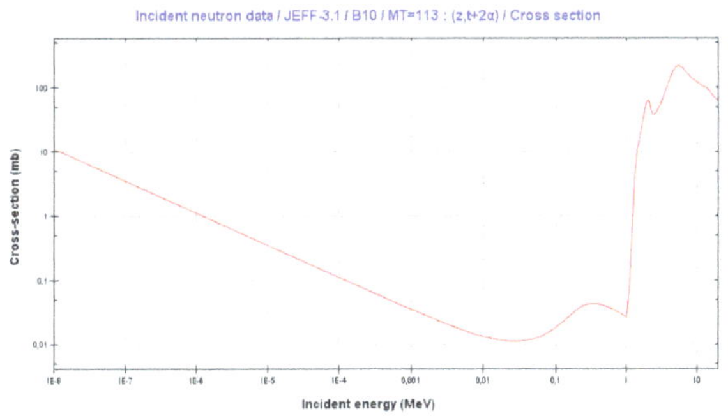

Cross-section of 10B(n,2alpha)T reaction

Tritium is a byproduct in nuclear reactors. Most of the tritium produced in nuclear power plants stems from the boric acid, which is commonly used as a chemical shim to compensate an excess of initial reactivity. Main reaction, in which the tritium is generated from boron is below:

10B(n,2*alpha)T

This reaction of a neutron with an isotope ^{10}B is the main way, how radioactive tritium in primary circuit of all PWRs is generated. Note that, this reaction is a threshold reaction due to its cross-section.

Calculate the reaction Q-value.

The atom masses of the reactants and products are:

$$\begin{aligned} m(^{10}B) &= 10.01294\ amu \\ m(^{1}n) &= 1.00866\ amu \\ m(^{3}T) &= 3.01604\ amu \\ m(^{4}He) &= 4.0026\ amu \end{aligned}$$

Using the mass-energy equivalence, we get the Q-value of this reaction as:

Q = {(10.0129+1.00866) [amu] – (3.01604+2 x 4.0026) [amu]} x 931.481 [MeV/amu]

= 0.00036 x 931.481 = 0.335 MeV

Q-value of Endothermic Reactions

Example: Endothermic Reaction - Photoneutrons

In nuclear reactors the gamma radiation plays a significant role also in reactor kinetics and in a subcriticality control. Especially in nuclear reactors with D_2O moderator (CANDU reactors) or with Be reflectors (some experimental reactors). Neutrons can be produced also in (γ, n) reactions and therefore they are usually referred to as photoneutrons.

A high energy photon (gamma ray) can under certain conditions eject a neutron from a nucleus. It occurs when its energy exceeds the binding energy of the neutron in the nucleus. Most nuclei have binding energies in excess of 6 MeV, which is above the energy of most gamma rays from fission. On the other hand there are few nuclei with sufficiently low binding energy to be of practical interest. These are: ^{2}D, ^{9}Be, ^{6}Li, ^{7}Li and ^{13}C. As can be seen from the table the lowest threshold have ^{9}Be with 1.666 MeV and ^{2}D with 2.226 MeV.

Nuclide	Threshold(MeV)	Reaction
^{2}D	2.225	^{2}H(γ,n)^1 H
^{6}Li	3.697	^{6}Li(γ,n+p)^4 He
^{6}Li	5.67	^{6}Li(γ,n)^{5}Li
^{7}Li	7.251	^{7}Li(γ,n)^{6}Li
^{9}Be	1.667	^{9}Be(γ,n) ^{8}Be
^{13}C	4.9	^{13}C(γ,n)^{12}C

In case of deuterium, neutrons can be produced by the interaction of gamma rays (with a minimum energy of 2.22 MeV) with deuterium:

$$^2_1D+\gamma \rightarrow {}^1_1H+n$$

The reaction Q-value is calculated below:

The atom masses of the reactant and products are:

$$m(^2D) = 2.01363\ amu$$
$$m(^1n) = 1.00866\ amu$$
$$m(^1H) = 1.00728\ amu$$

Using the mass-energy equivalence, we get the Q-value of this reaction as:

Q = {2.01363 [amu] – (1.00866+1.00728) [amu]} x 931.481 [MeV/amu]

= -0.00231 x 931.481 = -2.15 MeV

Example: Endothermic Reaction - (α,n) reaction

Calculate the reaction Q-value of the following reaction:

7Li (α, n) 10B

The atom masses of the reactants and products are:

$$m(^4He) = 4.0026\ amu$$
$$m(^7Li) = 7.0160\ amu$$
$$m(^1n) = 1.0087\ amu$$
$$m(^{10}B) = 10.01294\ amu$$

Using the mass-energy equivalence, we get the Q-value of this reaction as:

Q = {(7.0160+4.0026) [amu] – (1.0087+10.01294) [amu]} x 931.481 [MeV/amu]

= 0.00304 x 931.481 = -2.83 MeV

Nuclide Stability

Nuclear stability means that nucleus is stable meaning that it does not spontaneously emit any kind of radioactivity (radiation). On the other hand, if the nucleus is unstable (not stable), it has the tendency of emitting some kind of radiation, i.e., it is radioactive. Therefore the radioactivity is associated with unstable nucleus:

Stable nucleus – non-radioactive

Unstable nucleus – radioactive

Keep in mind that less stable means more radioactive and more stable means less radioactive.

Features of Stable Nuclides

There are several features with respect to stable nuclides.

- Numbers of protons and neutrons

  Only ^{1}H and ^{3}He have more protons than neutrons.

  Nuclides ^{2}D, ^{4}He, ^{6}Li, ^{10}B, ^{12}C, ^{14}N, ^{16}O, ^{20}Ne, ^{24}Mg, ^{28}Si, ^{32}S, ^{36}Ar, and ^{40}Ca have equal number of protons and neutrons. All other nuclides have more neutrons than protons.

  Only ^{2}D, ^{6}Li, ^{10}B, and ^{14}N have equal but odd number of protons and neutrons.

  Heavy stable nuclides have more neutrons per proton than light ones have.

- Magic numbers of protons and neutrons

  Elements with Z = 2, 8, 20, 28, 50, and 82 have many isotopes. Thus, these numbers are called magic numbers. For example, the stable isotopes of calcium (*Z* = 20) have mass numbers 40, 42, 43, 44, 46, and 48, a total of six. There are a total of ten stable isotopes of tin, $^{112, 114, 115, 116, 117, 118, 119, 120, 122, \& 124}Sn^{50}$.

  In our discussion of radioactive decay families, you have noticed that the stable element of three families is lead (*Z* = 82), and it has a magic number of protons. The stable nuclide for the fourth family is $^{209}Bi^{83}$, and it has a magic number (126) of neutrons. This is the stable nuclide with the highest atomic number.

- Double-magic-number nuclides

  Nuclides ^{4}He, $^{16}O^{8}$, $^{40}Ca^{20}$, $^{48}Ca^{20}$, and $^{208}Pb^{82}$ have magic numbers of protons and magic number of neutrons.

- Pairing of nucleons

Since nucleons have ½ spin, they obey Pauli's exclusion principle by allowing two protons or neutrons each with opposite spin to occupy a quantum state (if they are nucleons in a nucleus). There is a preference for having pairs of protons or neutrons, and it is known as pairing of nucleons.

Effect of Paring Nucleons		
Z	*N*	No. of stable nuclides
even	even	166
even	odd	57
odd	even	53
odd	odd	4

More than half (59%) of stable nuclides have even numbers of protons and neutrons. This fact suggests that pairing of protons and neutrons contributes to the stability of nuclides.

The effect of pairing also affects the abundance of the isotopes in elements, as well as the abundance of a nuclide on a planet, galactic or universal scale.

Nucleon pairing also affects the decay of unstable nuclides.

- Abundance of elements

The abundance of an element or nuclide is its amount in a system. Elemental abundances of stars and galaxies can be determined from X-ray spectra in space explorations.

The sun has 99.9% of the mass of the solar system. Hydrogen atoms contribute 72%, and helium 4He 26% to all atoms in the Sun.

Taking as a whole, the most abundant element of the planet Earth is iron, which is the major component of the earth (molten) core. Additional evidence comes from the many iron meteorites, which are considered debris from outer space. However, the most abundant element of the Earth crust is oxygen in terms of number of atoms, but silicon is the most abundant element by mass.

Example

Based on the even-odd rule presented above, predict which one would you expect to be radioactive in each pair?

(a) $^{16}_{8}O$ and $^{17}_{8}O$

(b) $^{35}_{17}\mathrm{Cl}$ and $^{36}_{17}\mathrm{Cl}$

(c) $^{20}_{10}\mathrm{Ne}$ and $^{17}_{10}\mathrm{Ne}$

(d) $^{40}_{20}\mathrm{Ca}$ and $^{45}_{20}\mathrm{Ca}$

(e) $^{195}_{80}Hg$ and $^{196}_{80}Hg$

Answer

(a) The $^{16}_{8}O$ contains 8 protons and 8 neutrons (even-even) and the $^{17}_{8}O$ contains 8 protons and 9 neutrons (even-odd). Therefore, $^{17}_{8}O$ is radioactive.

(b) The $^{35}_{17}Cl$ has 17 protons and 18 neutrons (odd-even) and the $^{36}_{17}Cl$ has 17 protons and 19 neutrons (odd-odd). Hence, $^{36}_{17}Cl$ is radioactive.

(c) The $^{20}_{10}Ne$ contains 10 protons and 10 neutrons (even-even) and the $^{17}_{10}Ne$ contains 10 protons and 7 neutrons (even-odd). Therefore, $^{17}_{10}Ne$ is radioactive.

(d) The $^{40}_{20}Ca$ has even-even situation and $^{45}_{20}Ca$ has even-odd situation. Thus, $^{45}_{20}Ca$ is radioactive.

(e) The $^{195}_{80}Hg$ has even number of protons and odd number of neutrons and the $^{196}_{80}Hg$ has even number of protons and even number of neutrons. Therefore, $^{196}_{80}Hg$ is radioactive.

Valley of Nuclear Stability

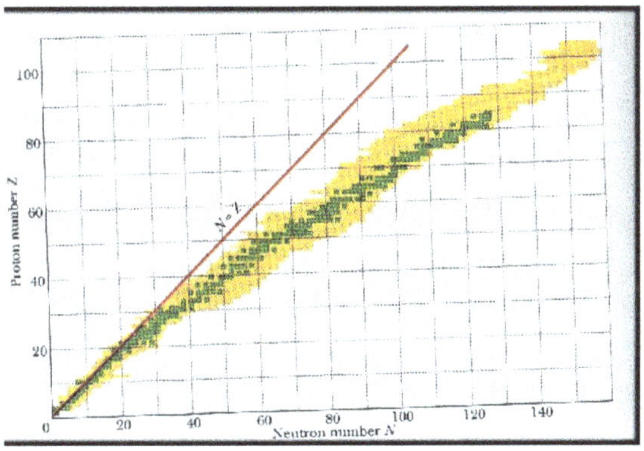

Number of protons and neutrons in nuclei

green square = stable nucleus

yellow square = radioactive nucleus

no square = no known nucleus

Examples:

Positive Beta Decay

Nuclei, such as ^{15}O, which are lacking in neutrons (consist of 8 protons and 7 neutrons) undergo positron decay (positive beta decay). In this process, one of the protons in the nucleus is transformed into a neutron, positron and neutrino. The positron and the neutrino are emitted. The number of protons is thus reduced from 8 to 7 (number of neutrons is increased from 7 to 8), so that the resulting nucleus is an isotope of nitrogen, ^{15}N, which is stable.

Negative Beta Decay

On the other hand nuclei, such as ^{19}O, which have excess of neutrons, decay by negative beta decay, emitting a negative electron and an antineutrino. In this process, one of the neutrons in the nucleus is transformed into a proton. The number of protons is thus increased from 8 to 9 (number of neutrons is reduced from 11 to 10), so that the resulting nucleus is an isotope of fluor, ^{19}F, which is stable. It should be noted that in both positive or negative beta decays the atomic mass number remains the same.

Nuclear Stability – Periodic Table

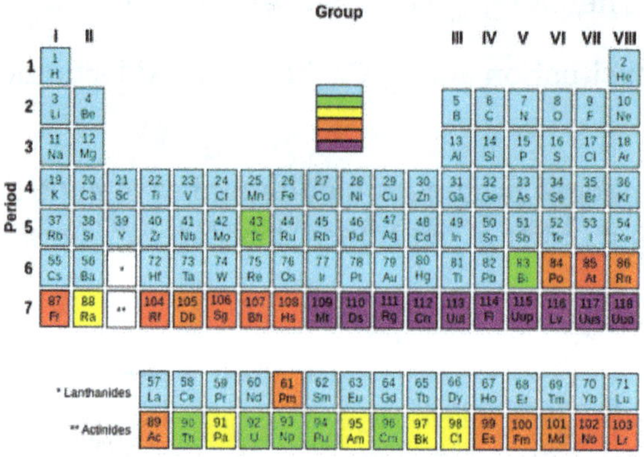

Periodic table with elements colored according to the half-life of their most stable isotope.

Of the first 82 elements in the periodic table, 80 have isotopes considered to be stable. Technetium, promethium and all the elements with an atomic number over 82 are unstable and decompose through radioactive decay. No undiscovered heavy elements (with atomic number over 110) are expected to be stable, therefore lead is considered the heaviest stable element. For each of the 80 stable elements, the number of the stable isotopes is given. For example, tin has 10 such stable isotopes.

There are 80 elements with at least one stable isotope, but 114 to 118 chemical elements are known. All elements to element 98 are found in nature, and the remainder of the discovered elements are artificially produced, with isotopes all known to be highly radioactive with relatively short half-lives.

Bismuth, thorium, uranium and plutonium are primordial nuclides because they have half-lives long enough to still be found on the Earth, while all the others are produced either by radioactive decay or are synthesized in laboratories and nuclear reactors. Primordial nuclides are nuclides found on the Earth that have existed in their current form since before Earth was formed. Primordial nuclides are residues from the Big Bang, from cosmogenic sources, and from ancient supernova explosions which occurred before the formation of the solar system. Only 288 such nuclides are known.

Connection between Nuclear Stability and Radioactive Decay

The nuclei of radioisotopes are unstable. In an attempt to reach a more stable arrangement of its neutrons and protons, the unstable nucleus will spontaneously decay to form a different nucleus. If the number of neutrons changes in the process (number of protons remains), a different isotopes is formed and an element remains (e.g. neutron emission). If the number of protons changes (different atomic number) in the process, then an atom of a different element is formed. This decomposition of the nucleus is referred to as radioactive decay. During radioactive decay an unstable nucleus spontaneosly and randomly decomposes to form a different nucleus (or a different energy state – gamma decay), giving off radiation in the form of atomic partices or high energy rays. This decay occurs at a constant, predictable rate that is referred to as half-life. A stable nucleus will not undergo this kind of decay and is thus non-radioactive.

References

- Bodanis, David (2009). E=mc^2: A Biography of the World's Most Famous Equation(illustrated ed.). Bloomsbury Publishing. ISBN 978-0-8027-1821-1
- What-is-nuclear-energy: nnr.co.za, Retrieved 16 April 2018
- Hecht, Eugene (2011), "How Einstein confirmed E0=mc2", American Journal of Physics, 79 (6): 591–600, Bibcode:2011AmJPh..79..591H, doi:10.1119/1.3549223
- Dr. Rod Nave of the Department of Physics and Astronomy, Dr. Rod Nave (July 2010). "Nuclear Binding Energy". Hyperphysics - a free web resource from GSU. Georgia State University. Retrieved 2010-07-11
- Nuclear-energy: whatisnuclear.com, Retrieved 28 June 2018
- Fewell, M. P. (1995). "The atomic nuclide with the highest mean binding energy". American Journal of Physics. 63 (7): 653–658. Bibcode:1995AmJPh..63..653F. doi:10.1119/1.17828
- Jammer, Max (1997) [1961], Concepts of Mass in Classical and Modern Physics, New York: Dover, ISBN 0-486-29998-8
- Nuclear-binding-energy, physical-chemistry: tutorvista.com, Retrieved 16 March 2018
- Lewis, Gilbert N. & Tolman, Richard C. (1909), "The Principle of Relativity, and Non-Newtonian Mechanics", Proceedings of the American Academy of Arts and Sciences, 44(25): 709–726, doi:10.2307/20022495
- "Nuclear binding energy". How to solve for nuclear binding energy. Guides to solving many of the types of quantitative problems found in Chemistry 116. Purdue University. July 2010. Retrieved 2010-07-10
- Weizsaecker-formula-semi-empirical-mass-formula, nuclear-power: nuclear-power.net, Retrieved 11 May 2018
- Prentiss, J.J. (August 2005), "Why is the energy of motion proportional to the square of the velocity?", American Journal of Physics, 73 (8): 705, Bibcode:2005AmJPh..73..701P, doi:10.1119/1.1927550

4

Interaction of Particles with Matter

All the diverse interactions of particles with matter and the different forces acting on these have been covered in this chapter. The topics covered herein include interaction of proton with matter, interaction of neutron with matter, interaction of heavy charged particles with matter, weak nuclear force, strong nuclear force, etc.

Nuclear Interaction

Nuclear forces (also known as nuclear interactions or strong forces) are the forces that act between two or more nucleons. They bind protons and neutrons ("nucleons") into atomic nuclei. The nuclear force is about 10 millions times stronger than the chemical binding that holds atoms together in molecules. This is the reason why nuclear reactors produce about a million times more energy per kilogram fuel as compared to chemical fuel like oil or coal. However, the range of the nuclear force is short, only a few femtometer (1 fm $=10^{-15}$ m), beyond which it decreases rapidly. That is why, in spite of its enormous strength, we do not feel anything of this force on the atomic scale or in everyday life. The development of a proper theory of nuclear forces has occupied the minds of some of the brightest physicists for seven decades and has been one of the main topics of physics research in the 20th century. The original idea was that the force is caused by the exchange of particles lighter than nucleons known as mesons, and this idea gave rise to the birth of a new sub-field of modern physics, namely, (elementary) particle physics. The modern perception of the nuclear force is that it is a residual interaction (similar to the van der Waals force between neutral atoms) of the eFigureven stronger force between quarks, which is mediated by the exchange of gluons and holds the quarks together inside a nucleon.

Properties of the Nuclear Force

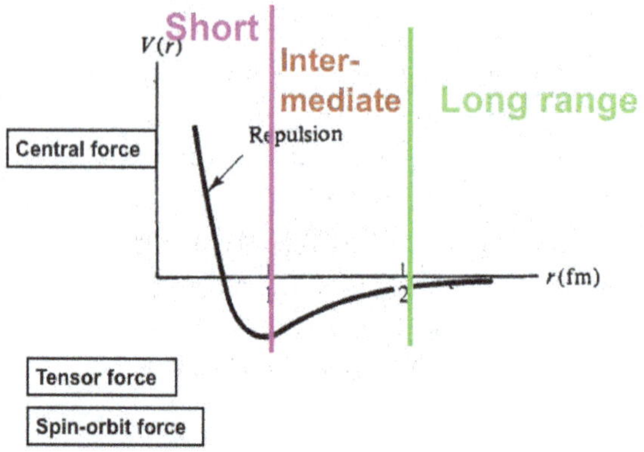

Figure: The major components of the nuclear force. The central force is best analyzed in terms of the three ranges indicated.

Some properties of nuclear interactions can be deduced from the properties of nuclei. Nuclei exhibit a phenomenon known as saturation: the volume of nuclei increases proportionally to the number of nucleons. This property suggests that the nuclear (central) force is of short range (a few fm) and strongly attractive at that range, which explains nuclear binding. But the nuclear force has also a very complex spin-dependence. Evidence of this property first came from the observation that the deuteron (the proton-neutron bound state, the smallest atomic nucleus) deviates slightly from a spherical shape: it has a non-vanishing quadrupole moment. This suggests a force that depends on the orientation of the spins of the nucleons with regard to the vector joining the two nucleons (a tensor force). In heavier nuclei, a shell structure has been observed which, according to a proposal by M. G. Mayer and J. H. D. Jensen, can be explained by a strong force between the spin of the nucleon and its orbital motion (the spin-orbit force). More clear-cut evidence for the spin-dependence is extracted from scattering experiments where one nucleon is scattered off another nucleon, with distinct spin orientations. In such experiments, the existence of the nuclear spin-orbit and tensor forces has clearly been established. Scattering experiments at higher energies (more than 200 MeV) provide evidence that the nucleon-nucleon interaction turns repulsive at short inter-nucleon distances (smaller than 0.5 fm, the hard core).

Besides the force between two nucleons (2NF), there are also three-nucleon forces (3NFs), four-nucleon forces (4NFs), and so on. However, the 2NF is much stronger than the 3NF, which in turn is much stronger than the 4NF, and so on. In exact calculations of the properties of light nuclei based upon the "elementary" nuclear forces, it has been shown that 3NFs are important. Their contribution is small, but crucial. The need for 4NFs for explaining nuclear properties has not (yet) been established.

Nuclear forces are approximately charge-independent meaning that the force between two protons, two neutrons, and a proton and a neutron are nearly the same (in the same quantum mechanical state) when electromagnetic forces are ignored.

Phenomenological Approaches

Traditionally, research on the nuclear force has proceeded along two lines: phenomenology and theory. We start with phenomenology. At very low energy, nucleon-nucleon scattering can be described by the so-called effective range expansion which, in its simplest version, has only two parameters. In modern times, this expansion has been recovered from an EFT which does not include the pion. At intermediate energies, the spin-dependence of the nuclear force becomes important. To produce a general expression for the *NN* potential (that includes spin and momentum dependences), imposed the following symmetries,

- Translational invariance
- Galilean invariance
- Rotational invariance
- Space reflection invariance
- Time reversal invariance
- Invariance under the interchange of particle 1 and 2
- Isospin symmetry

- **Hermiticity**

and obtained:

$$\begin{aligned} V = {} & V_C + \tau_1.\tau_2\, W_C && \textit{central} \\ & + [V_S + \tau_1.\tau_2\, W_S]\vec{\sigma}_1.\vec{\sigma}_2 && \textit{spin-spin} \\ & + [V_{LS} + \tau_1.\tau_2\, W_{LS}]\vec{L}.\vec{S} && \textit{spin-orbit} \\ & + [V_T + \tau_1.\tau_2\, W_T]S_{12}(\vec{r}) && \textit{tensor} \\ & + [V_{\sigma L} + \tau_1.\tau_2\, W_{\sigma L}]Q_{12} && \sigma - L \\ & + [V_{\sigma p} + \tau_1.\tau_2\, W_{\sigma p}]\vec{\sigma}_1.\vec{p}.\vec{\sigma}_2.\vec{p} && \sigma - p \end{aligned}$$

with

$$S_{12}(\vec{r}) \equiv 3\vec{\sigma}_1.\hat{r}.\vec{\sigma}_2.\hat{r} - \vec{\sigma}_1.\vec{\sigma}_2,$$

$$Q_{12} \equiv \frac{1}{2}\left[\vec{\sigma}_1.\vec{L}\vec{\sigma}_2.\vec{L} + \vec{\sigma}_2.\vec{L}\,\vec{\sigma}_1.\vec{L}\right],$$

and

$\vec{r} \equiv \vec{r}_1 - \vec{r}_2$ relative coordinate,

$\hat{r} \equiv \vec{r}/r$ unit vector for relative coordinate,

$\vec{p} \equiv \frac{1}{2}(\vec{p}_1 - \vec{p}_2)$ relative momentum,

$\vec{L} \equiv \vec{L}_1 + \vec{L}_2 = \vec{r}\times\vec{p} = -i\vec{r}\times\vec{\nabla}$ total orbital angular momentum in position space,

$\vec{S} \equiv \frac{1}{2}(\vec{\sigma}_1 + \vec{\sigma}_2)$ total spin,

where $\vec{r}_{1,2}$ $\vec{p}_{1,2}$, $\vec{L}_{1,2}$, $\vec{\sigma}_{1,2}$, and $\boldsymbol{\tau}_{1,2}$ denote position, momentum, angular momentum, spin, and isospin, respectively, of nucleon 1 and 2. The *Vi* and *Wi*, with *i=C, S, LS, T, σL, σp,* are functions of r^2, p^2, and L^2 only, i.e.

$$V_i = V_i(r^2, p^2, L^2),$$
$$W_i = W_i(r^2, p^2, L^2).$$

Charge-independence or isospin invariance requires that the potential is a scalar in the isospin space of the two nucleons. The only such scalars are 1 and $\boldsymbol{\tau}_1\cdot\boldsymbol{\tau}_2$, which explains the isospin structures in

$$\text{Equation}\quad \begin{aligned} V = {} & V_C + \tau_1.\tau_2\, W_C && \textit{central} \\ & + [V_S + \tau_1.\tau_2\, W_S]\vec{\sigma}_1.\vec{\sigma}_2 && \textit{spin-spin} \\ & + [V_{LS} + \tau_1.\tau_2\, W_{LS}]\vec{L}.\vec{S} && \textit{spin-orbit}\,. \\ & + [V_T + \tau_1.\tau_2\, W_T]S_{12}(\vec{r}) && \textit{tensor} \end{aligned}$$

$$+\left[V_{\sigma L}+\tau_1.\tau_2\,W_{\sigma L}\right]Q_{12} \qquad \sigma-L$$
$$+\left[V_{\sigma p}+\tau_1.\tau_2\,W_{\sigma p}\right]\vec{\sigma}_1.\vec{p}.\vec{\sigma}_2.\vec{p} \quad \sigma-p$$

If energy is conserved in the scattering process ("on shell"), then there are only five independent terms and the *σp* term can be expressed as a combination of the other five terms. Note, however, that when a potential is applied in a scattering equation (Schrödinger or Lippmann-Schwinger equation) the potential goes off shell.

Potentials which are based upon the operator structure in above equation with functions V_i and W_i chosen such as to fit the *NN* data or phase shifts are called phenomenological potentials. To keep things simple, most phenomenological potentials do not include all six terms. A minimal set for a realistic potential is the central, spin-spin, spin-orbit and tensor term.

NN potentials are:

- *Gammel and Thaler*, first semi-quantitative *NN* potential, hard-core.
- *Hamada and Johnston*, first quantitative *NN* potential, hard core.
- *Reid*, first quantitative soft core potential, very popular in the 1970s.
- *Argonne V14 potential*, based upon a set of 14 operators.
- *Argonne V18 potential*, based upon a set of 18 operators, charge.

Meson Theory of Nuclear Forces

Yukawa's Idea of 1935

In 1935, Yukawa introduced the concept of massive particle exchange to explain the nuclear force. He constructed an analogy to classical electrodynamics. In electrodynamics, the Coulomb potential

$$\phi(r)=\frac{q}{4\pi}\frac{1}{r}$$

is the solution of Poisson's equation

$$\nabla^2\phi(\vec{r}\,)=-q\delta^{(3)}(\vec{r}\,).$$

When adding a mass term to this equation (and flipping the sign on the r.h.s. to adjust for scalar coupling),

$$(\nabla^2-m^2)\,\varphi(\vec{r})=g\delta^{(3)}(\vec{r}),$$

the solution becomes

$$\varphi(r)=-\frac{g}{4\pi}\frac{e^{-mr}}{r},$$

which is the scalar field generated by one nucleon. A second nucleon, with also coupling g, at a distance r from the first one will be exposed to the interaction energy

$$V(r) = -\frac{g^2}{4\pi}\frac{e^{-mr}}{r},$$

which is known as the Yukawa potential. The exponential in this expression, that is due to the mass m of the meson, restricts the potential to a finite range, which is the essential point. For $m \rightarrow 0$ we are back to the form of the Coulomb potential.

One-boson-exchange Model

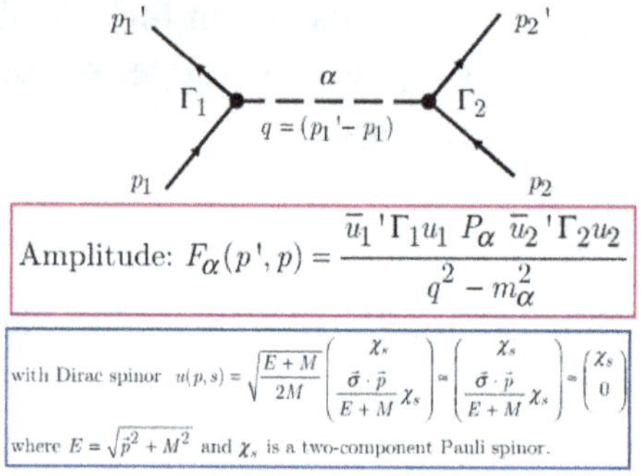

Figure: Feynman diagram describing the exchange of a boson α between two nucleons. The amplitude $F\alpha$ that results from this diagram is stated, too. The $P\alpha$ and the Γi for the various meson exchanges are given in below Figures.

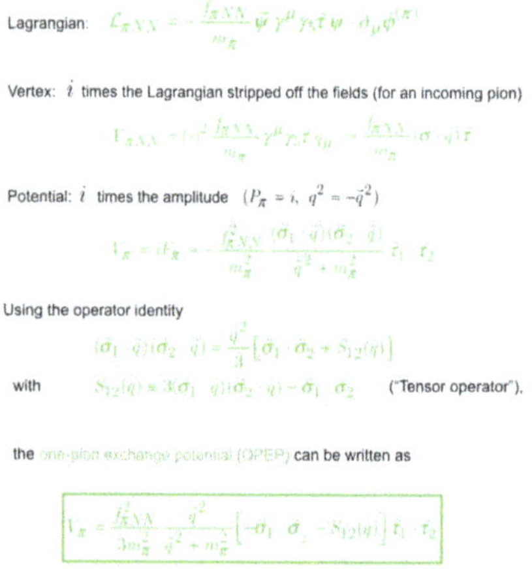

Figure: Derivation of the one-pion-exchange potential in non-relativistic approximation.

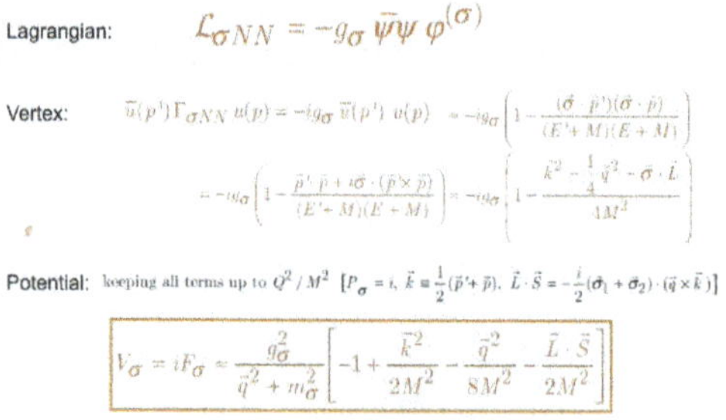

Figure: Derivation of the one-sigma-exchange potential up to terms in $Q2/M2$.

Lagrangian: $\mathcal{L}_{\omega NN} = -g_\omega \bar{\psi}\gamma^\mu \psi \, \varphi_\mu^{(\omega)}$

Vertex:

$$\mu = 0: \; \bar{u}(p')\Gamma^0_{\omega NN} u(p) = -ig_\omega \bar{u}(p')\gamma^0 u(p) \approx -ig_\omega \left(1 + \frac{(\vec{\sigma}\cdot\vec{p}\,')(\vec{\sigma}\cdot\vec{p})}{(E'+M)(E+M)}\right)$$

$$\approx -ig_\omega \left(1 - \frac{\vec{\sigma}\cdot\vec{L}}{4M^2}\right), \text{ keeping only the } \vec{\sigma}\cdot\vec{L} \text{ term.}$$

Potential, including also the $\vec{\gamma}$ terms: $\left[P_\omega = -ig_{\mu\nu} + \ldots\right]$

$$\boxed{V_\omega = iF_\omega \approx \frac{g_\omega^2}{\vec{q}^{\,2} + m_\omega^2}\left[1 - 3\frac{\vec{L}\cdot\vec{S}}{2M^2}\right]}$$

Figure: Derivation of the one-omega-exchange potential keeping only the central and spin-orbit term.

Yukawa's original derivation was done for scalar bosons. When finally a real meson was discovered in 1947/48, it turned out to be pseudo-scalar with mass around 138 MeV and was dubbed the π-meson or pion. Consequently, in the 1950s, the attempts to derive the nuclear force focused on theories that included only pions. These 'pion theories' had many problems and little success--for reasons we understand today: pion dynamics is constrained by chiral symmetry, a concept that was unknown in the 1950s. In the early 1960s, heavier (non-strange) mesons were found in experiment, notably the vector (spin-1) mesons $\rho(770)$ and $\omega(782)$. Because of the problems with the pion theories, theoreticians were now happy to extend meson theory by including more and different species of mesons. This led to the one-boson-exchange (OBE) models, which were started in the 1960s and turned out to be very successful for the two-nucleon interaction.

Let's first address the question of which mesons to consider. When deriving the nuclear force, one has generally more confidence in the predictions for the longer ranged parts. Since the range, *Ra*, of the force created by a meson is inversely proportional to the meson mass, *ma*, i.e.,

$$R_a \sim \frac{1}{m_a},$$

one starts with the lightest mesons and moves up to mesons with masses in the order of the nucleon mass. This includes essentially six mesons, namely, $\pi(138)$, $\eta(548)$, $\sigma(500)$, $\rho(770)$, $\omega(782)$, *a*0(980), where the numbers in parentheses are the masses in MeV. As it turns out, η and *a*0 are not very important and, so, we will focus here on

- the pseudo-scalar isovector pion (0−,1),
- the scalar isoscalar sigma (0+,0),
- the vector isoscalar omega (1−,0),
- the vector isovector rho (1−,1),

where the parenthetical information, (*JP*,*I*), summarizes spin *J*, parity *P*, and isospin *I* for each particle.

Yukawa's original considerations used classical field theory. A more proper derivation should be based upon quantum field theory, as we will use here. In the one-boson-exchange (OBE) model, the mesons are exchanged singly as shown in the Feynman diagram of above Figure. The contributions to the *NN* potential from the various mesons are derived in above Figures. In these derivations, we always start from an appropriate interaction Lagrangian for meson-nucleon coupling, which is designed with guidance from symmetry principles (the Lagrangian must be a Lorentz scalar). Concerning the Lagrangians for the vector mesons ω and ρ, we note that they may have both a vector and a tensor coupling (with coupling constants gv and fv, respectively):

$$\mathcal{L}_{vNN} = -g_v\bar{\psi}\gamma^{\mu}\psi\varphi_{\mu}^{(v)} - \frac{f_v}{4M}\bar{\psi}\sigma^{\mu\nu}\psi(\partial_{\mu}\varphi_{\nu}^{(v)} - \partial v\varphi_{\mu}^{(v)}).$$

These two couplings are similar to the interaction of a photon with a nucleon. The first is analogous to the coupling of the Dirac current to the electromagnetic vector potential, while the second one corresponds to the Pauli coupling of the anomalous magnetic moment. The analogy is not accidental; the vector-meson dominance model (VDM) for the electromagnetic form factor of the nucleon explains the close relationship. In the VDM one assumes that the photon couples to the nucleon through a vector boson, which explains the extended structure of the nucleon electromagnetic form factor. In the strict interpretation of this model, the ρ coupling constants ratio, $f\rho/g\rho$, should be 3.7, and from dispersion analysis one obtains even $f\rho/g\rho \approx 6$. In any case, the tensor coupling of the ρ is much larger than its vector coupling, which is why we omitted the vector coupling in above Figure. For the ω meson, it is the other way around: the vector dominance model suggests a ω coupling constants ratio $f\omega/g\omega = -0.12$. Since this is close to zero, the ω is given no tensor coupling in most meson models.

The full propagator for vector bosons is

$$P_v = i\frac{-g_{\mu\nu} + q_{\mu}q_{\nu}/m_v^2}{q^2 - m_v^2},$$

where in above figure we dropped the $q_{\mu}q_{\nu}/m_v^2$ – term. Due to nuclear current conservation, this term vanishes on-shell, but not off-shell. The off-shell effect of this term was examined by Holinde and Machleidt was found to be unimportant.

It is customary to multiply the vertices with cutoffs, which suppress high-momentum components to ensure the convergence of the scattering equation. A simple form for these cutoffs is

$$\left(\frac{\Lambda_{\alpha}^2 - m_{\alpha}^2}{\Lambda_{\alpha}^2 + \vec{q}^{\,2}}\right)^{n_{\alpha}},$$

Where the cutoff mass $\Lambda\alpha$ is typically chosen in the range 1.3 - 2.0 GeV. The multiplication by these form factors is not explicitly shown in our derivations. The calculation is performed in momentum space and in the center-of-mass (CMS) system of the two interacting nucleons,

where $p_1 = (E, \vec{p})$ and $p_2 = (E, -\vec{p})$ in the initial states; and $p'_1 = (E', \vec{p}')$ and $p'_2 = (E', -\vec{p}')$ in the final states. Moreover,

$\vec{q} \equiv \vec{p}' - \vec{p}$ is the momentum transfer,

$\vec{k} \equiv \frac{1}{2}(\vec{p}' + \vec{p})$ is the momentum transfer,

$\vec{L} = -i(\vec{q} \times \vec{k}) = -i(\vec{p}' + \vec{p})$ the total orbital angular momentum in momentum space,

$S_{12}(\vec{q}) \equiv 3\vec{\sigma}_1.\hat{q}.\vec{\sigma}_2.\hat{q} - \vec{\sigma}_1.\vec{\sigma}_2$ the tensor operator in momentum space.

The characteristic properties of the contributions from the various mesons derived in above figures are summarized in below table.

The OBE *NN* potential (OBEP) is defined as the sum over the contributions from the four mesons discussed and, typically, a few more, e.g.,

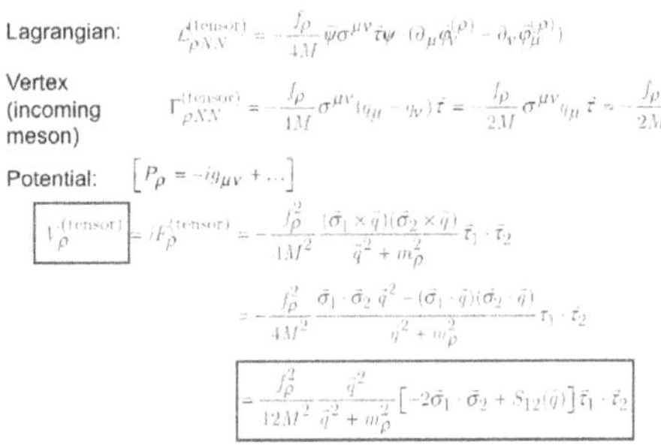

Figure: Derivation of the one-rho-exchange potential in non-relativistic approximation using only the tensor coupling of the ρ to the nucleon.

$$V_{\text{OBE}} = \sum_{a=\pi,\sigma,\omega,\rho,\eta,a_0} V_a.$$

Meson	Central	Spin-Spin	Tensor	Spin-Orbit
π(138)	---	weak, long-ranged	strong, long-ranged	---
σ(500)	strong, attractive, intermediate-ranged	---	---	moderate, intermediate-ranged
ω(782)	strong, repulsive, short-ranged	---	---	strong, short-ranged, coherent with σ
ρ(770)	---	weak, short-ranged, coherent with π	moderate, short-ranged, opposite to π	---

Table: The four most important mesons and the main characteristics of their contributions to components of the nuclear force.

OBEP in Position Space

The momentum-space potentials can be Fourier transformed,

$$V_\alpha(\vec{r}) = \frac{1}{(2\pi)^3}\int d^3q\, e^{i\vec{q}.\vec{r}\, V_\alpha(\vec{q})},$$

to obtain the following equivalent position space potentials:

$$V_\pi(\vec{r}) = \frac{1}{3}\frac{f_{\pi NN}^2}{4\pi} m_\pi \left\{\left[Y(m_\pi r) - \frac{4\pi}{m_\pi^3}\delta^{(3)}(\vec{r})\right]\vec{\sigma}_1.\vec{\sigma}_2 + \left(1 + \frac{3}{m_\pi r} + \frac{3}{(m_\pi r)^2}\right) Y(m_\pi r) S_{12}(\hat{r})\right\}\tau_1.\tau_2,$$

$$V_\sigma(\vec{r}) = -\frac{g_\sigma^2}{4\pi} m_\sigma \left\{\left[1 - \frac{1}{4}\left(\frac{m_\sigma}{M}\right)^2\right] Y(m_\sigma r) + \frac{1}{4M^2}\left[\vec{\nabla}^2 Y(m_\sigma r) + Y(m_\sigma r)\vec{\nabla}^2\right]\right.$$

$$+\frac{1}{2}\left(\frac{m_\sigma}{M}\right)^2\left(\frac{1}{m_\sigma r}+\frac{1}{(m_\sigma r)^2}\right)Y(m_\sigma r)\vec{L}.\vec{S}\Bigg\},$$

$$V_w(\vec{r})=\frac{g_w^2}{4\pi}m_v\left[Y(m_\omega r)-\frac{3}{2}\left(\frac{m_w}{M}\right)^2\left(\frac{1}{m_\omega r}+\frac{1}{(m_\omega r)^2}\right)Y(m_\omega r)\vec{L}.\vec{S}\right],$$

$$V_\rho(\vec{r})=\frac{1}{12}\frac{f_\rho^2}{4\pi}\left(\frac{m_\rho}{M}\right)^2 m_\rho\left\{2\left[Y(m_\rho r)-\frac{4\pi}{m_\rho^3}\delta^{(3)}(\vec{r})\right]\vec{\sigma}_1.\vec{\sigma}_2+\left(1+\frac{3}{m_\rho r}+\frac{3}{(m_\rho r)^2}\right)Y(m_\rho r)S_{12}(\hat{r})\right\}\tau_1.\tau_2,$$

with the "Yukawa function"

$Y(x)=e^{-x}/x$

and $S12(\hat{r})$ given in Equation $S_{12}(\vec{r})\equiv 3\vec{\sigma}_1.\hat{r}.\vec{\sigma}_2.\hat{r}-\vec{\sigma}_1.\vec{\sigma}_2,$.

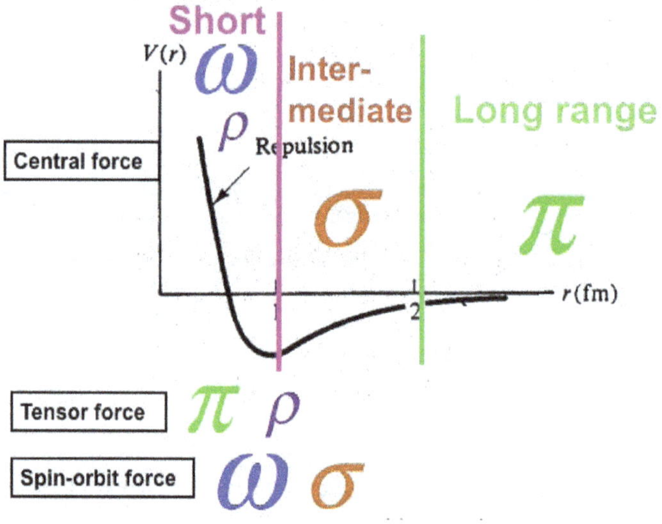

Figure: Cartoon of the nuclear force in the meson picture.

We can now better understand the success of the OBE model by summarizing its properties:

- The pseudo-scalar pion with a mass of about 138 MeV is the lightest meson and provides the longest-ranged part of the *NN* potential and the essential part of the tensor force.
- The vector meson $\boldsymbol{\rho}$ (770 MeV) cuts down the tensor force created by the pion at short distances to arrive at a tensor force of realistic strength.
- The scalar $\boldsymbol{\sigma}$ boson of about 500 MeV provides the intermediate-range attraction necessary for nuclear binding.
- The vector meson $\boldsymbol{\omega}$ (782 MeV) produces a strong repulsive central force of short range (repulsive core) and the essential part of the nuclear spin-orbit force.

This takes care of all the empirical properties of the nuclear force, therefore, allows for a *quantitative* description of the *NN* system.

Relativistic OBEPs

The first OBEPs ever developed were derived along the lines we followed above: one starts from the Feynman amplitude of an OBE and then performs a Q/M expansion up to terms of order $(Q/M)^2$. The motivation for this procedure was twofold: First, one wanted to see in a simple way what force components (e.g., central, spin-spin, tensor, spin-orbit) were created by the exchange of different mesons. Second, early researchers preferred a local potential in position space, i.e., an analytic expression for the potential that is a function of the relative distance between the two nucleons, $\vec{r}$. For this it is necessary that the Fourier transform of the momentum space expressions can be performed analytically.

However, there is no need to perform calculations in position space. Equally well and often in a more elegant way, calculations of *NN* scattering and nuclear bound states can be carried out in momentum space. Locality is not an issue and presents no advantages in momentum space. The original relativistic Feynman amplitudes of OBE are functions of p' and p and the relativistic OBEP is defined as:

$$V_{\mathrm{OBE}}^{\mathrm{rel}}(p',p) = i\sum_{\alpha} F_{\alpha}(p',p)$$

with $F\alpha(p',p)$ as in above figure, evaluated in full beauty and without approximations. Two-nucleon scattering is described covariantly by the Bethe-Salpeter equation, for which $V_{\mathrm{OBE}}^{\mathrm{rel}}$ is input. Since the four-dimensional Bethe-Salpeter (BS) equation is difficult to solve , relativistic three-dimensional (3D) reductions of the BS equation are frequently used, which are more amenable to numerical solution. It is common to the derivation of all relativistic three-dimensional equations that the time component of the relative momentum is fixed in some covariant way, so that it no longer appears as a separate variable. Thus,

$$V_{\mathrm{OBE}}^{\mathrm{rel}}(p',p) \mapsto V_{\mathrm{OBE}}^{\mathrm{rel\,3D}}(\vec{p}',\vec{p}).\,\mathrm{S}$$

Beyond the OBE Approximation

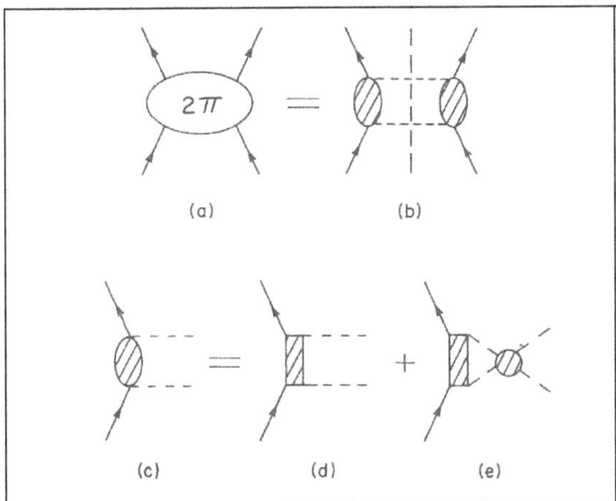

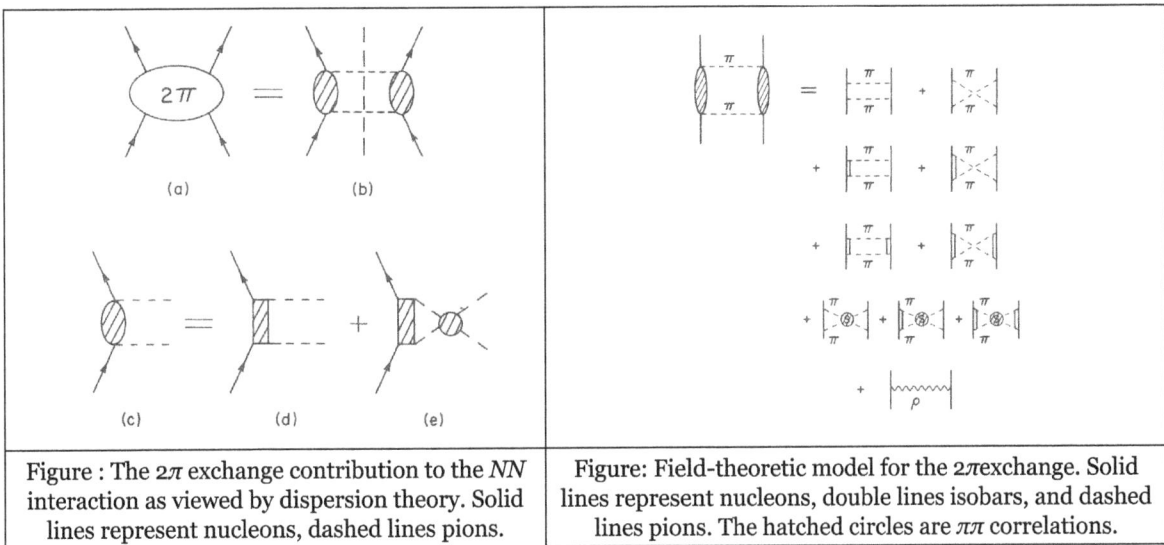

Figure : The 2π exchange contribution to the *NN* interaction as viewed by dispersion theory. Solid lines represent nucleons, dashed lines pions.

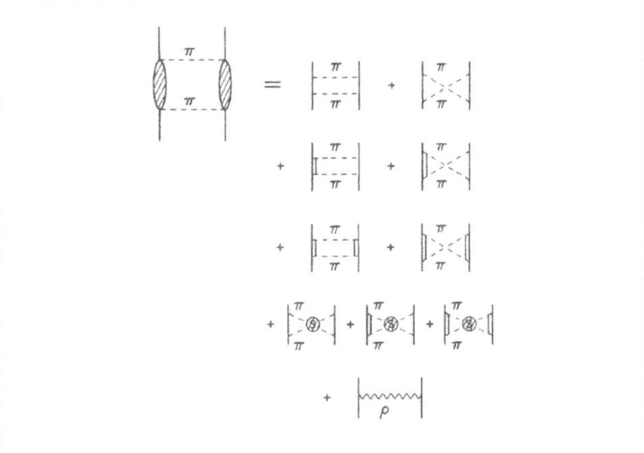

Figure: Field-theoretic model for the 2πexchange. Solid lines represent nucleons, double lines isobars, and dashed lines pions. The hatched circles are $\pi\pi$ correlations.

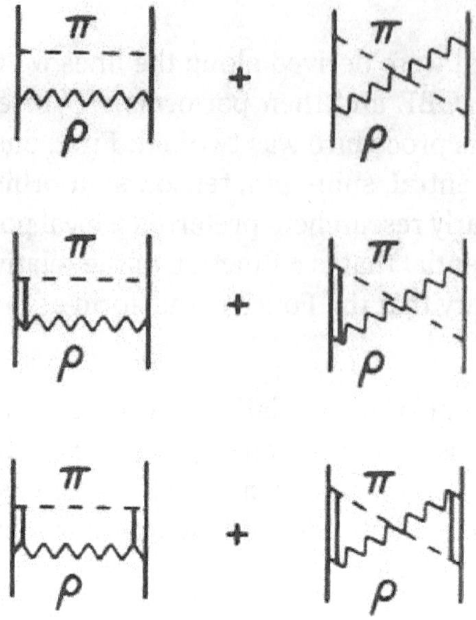

Figure: $\pi\rho$ contributions to the *NN*interaction.

The OBE model is a great simplification of the complicated scenario of a meson theory for the *NN* interaction. Therefore, in spite of the quantitative success of the OBEPs, we should be concerned about the approximations involved in the model. Major concerns include:

- The scalar isoscalar σ 'meson' of about 500 MeV.
- The neglect of all non-iterative diagrams.
- The role of meson-nucleon resonances.

Two pions, when 'in the air', can interact strongly. When in a relative *P*-wave (L=1), they form a proper resonance, the ρ meson. They can also interact in a relative *S*-wave (L=0), which gives rise to the σ boson. Whether the σ is a proper resonance is controversial, even though the Particle Data Group lists an f_0(500) or σ(500) meson, but with a width 400-700 MeV. What's for sure is that two pions have correlations, and if one doesn't believe in the σas a two pion resonance, then one has to take these correlations into account. There are essentially two approaches to take care of the two-pion exchange contribution to the *NN* interaction (which generates the intermediate range attraction): dispersion theory and field theory.

The *dispersion-theoretic* picture is described schematically in above figure. In this approach one assumes that the total diagram (a) can be analysed in terms of two 'halves' (b). The hatched ovals stand for all possible processes which two pions and a nucleon can undergo. This is made more explicit in (d) and (e). The hatched boxes represent all possible baryon intermediate states including the nucleon. (Note that there are also crossed exchanges which are not shown.) The shaded circle stands for $\pi\pi$ scattering. Quantitatively, these processes are taken into account by using empirical information from πN and $\pi\pi$scattering (e. g. phase shifts) which represents the input for such a calculation. Dispersion relations then provide an on-shell *NN* amplitude, which with some kind of plausible prescription is represented as a potential. The Paris potential is constructed along these lines complemented by one-pion-exchange and ω-exchange.

A *field-theoretic model* for the 2π-exchange contribution is shown in above figure. The model includes contributions from isobars as well as from $\pi\pi$correlations. This can be understood in analogy to the dispersion relations picture. In general, only the lowest-lying πN resonance, the so-called Δ isobar (spin 3/2, isospin 3/2, mass 1232 MeV), is taken into account. The contributions from other resonances have proven to be small for the low-energy *NN*processes under consideration. A field-theoretic model treats the Δ isobar as an elementary (Rarita-Schwinger) particle. The six upper diagrams of figure represent uncorrelated 2π exchange. The crossed (non-iterative) two-particle exchanges (second diagram in each row) are important. They guarantee the proper (very weak) isospin dependence due to characteristic cancellations in the isospin dependent parts of box and crossed box diagrams. Furthermore, their contribution is about as large as the one from the corresponding box diagrams (iterative diagrams); therefore, they are not negligible. In addition to the processes discussed, also correlated 2π exchange has to be included (lower two rows of above figure). Quantitatively, these contributions are about as large as those from the uncorrelated processes.

Besides the contributions from two pions, there are also contributions from the combination of other mesons. The combination of π and ρ is particularly significant, above figure. This contribution is repulsive and important to suppress the 2π exchange contribution at high momenta (or small distances), which is too strong by itself.

The most developed meson-theoretic model for the *NN* interaction is the Bonn Full Model, which includes all the diagrams displayed in above figures plus single π and ω exchange. Besides this, the Bonn Full Model also contains non-iterative graphs of π and σ and $\pi$and ω exchanges. However, the combined contribution of the latter two groups of diagrams is small and not significant.

Having highly sophisticated models at hand, like the Paris and the Bonn potentials, allows to check the approximations made in the simple OBE model. As it turns out, the highly complicated 2π exchange contributions to the *NN* interaction tamed by the $\pi\rho$ diagrams can well be simulated by the single scalar isoscalar boson, the σ, with a mass around 550 MeV. Retroactively, this fact provides justification for the simple OBE model.

Noteworthy Examples

Noteworthy examples for meson-theoretic *NN* potentials are:

- *Bryan and Scott* (1964, 1967, 1969), position-space OBEPs, among the first.
- *Early Bonn potentials*, relativistic momentum-space OBEPs.
- *Early Nijmegen potential*, position-space OBEP.
- *Paris potential*, position-space potential, 2π from dispersion theory plus π and ω.
- *Bonn Full Model*, most comprehensive meson-theoretic model for the *NN* interaction ever developed.
- *High-precision Nijmegen potential*, charge-dependent position-space OBEP, $\chi 2$/datum≈1 for fit of *NN* data.

- *High-precision Bonn potential (CD-Bonn,* charge-dependent momentum-space OBEP, $\chi 2$/datum≈1 for fit of *NN* data.
- *Gross and Stadler*, relativistic OBEP using the Gross equation, accurate fit of *np* data.

This could have been the happy end of the theory of nuclear forces. However, with the rise of QCD to the ranks of the authoritative theory of strong interactions, meson theory is demoted to the lower level of a model (even though a beautiful one).

QCD and the Nuclear Force

Quantum chromodynamics (QCD) is the theory of strong interactions. It deals with quarks, gluons and their interactions and is part of the Standard Model of Particle Physics. QCD is a non-Abelian gauge field theory with color *SU*(3) the underlying gauge group. The non-Abelian nature of the theory has dramatic consequences. While the interaction between colored objects is weak at short distances or high momentum transfer (asymptotic freedom), it is strong at long distances (≥1fm) or low energies, leading to the confinement of quarks into colorless objects, the hadrons. Consequently, QCD allows for a perturbative analysis at large energies, whereas it is highly non-perturbative in the low-energy regime. Nuclear physics resides at low energies and the force between nucleons is a residual color interaction similar to the van der Waals force between neutral molecules. Therefore, in terms of quarks and gluons, the nuclear force is a very complicated problem that, nevertheless, can be attacked with brute computing power on a discretized, Euclidean space-time lattice (known as lattice QCD). In a recent study , the neutron-proton scattering lengths in the singlet and triplet *S*-waves have been determined in fully dynamical lattice QCD, with a smallest pion mass of 354 MeV. This result is then extrapolated to the physical pion mass with the help of chiral perturbation theory. The pion mass of 354 MeV is still too large to allow for reliable extrapolations, but the feasibility has been demonstrated and more progress can be expected for the near future. In a lattice calculation of a very different kind, the nucleon-nucleon (*NN*) potential was studied. The central part of the potential shows a repulsive core plus attraction of intermediate range. This is a very promising result, but it must be noted that also in this investigation still rather large pion masses are being used. In any case, advanced lattice QCD calculations are under way and continuously improved. However, since these calculations are very time-consuming and expensive, they can only be used to check a few representative key-issues. For everyday nuclear structure physics, a more efficient approach is needed.

Chiral Effective Field Theory Approach to Nuclear Forces

The efficient approach is an effective field theory (EFT). For the development of an EFT, it is crucial to identify a separation of scales. In the hadron spectrum, a large gap between the masses of the pions and the masses of the vector mesons, like $\rho(770)$ and $\omega(782)$, can clearly be identified. Thus, it is natural to assume that the pion mass sets the "soft scale", $Q \sim m\pi$, and the rho mass the "hard scale", $\Lambda\chi \sim m\rho$, also known as the chiral-symmetry breaking scale. This is suggestive of considering an expansion in terms of the soft scale over the hard scale, $Q/\Lambda\chi$. Concerning the relevant degrees of freedom, we noticed already that, for the ground state and the low-energy excitation spectrum of an atomic nucleus as well as for conventional nuclear reactions, quarks and gluons are ineffective degrees of freedom, while nucleons and pions are the appropriate ones (this may include also low-lying nucleon resonances, s. below). To make sure that this EFT is not just another phenomenology, it must have a firm link with QCD. The link is established by having the EFT observe all relevant symmetries of the underlying theory. This requirement is based upon a "folk theorem" by Weinberg :

If one writes down the most general possible Lagrangian, including all terms consistent with assumed symmetry principles, and then calculates matrix elements with this Lagrangian to any given order of perturbation theory, the result will simply be the most general possible S-matrix consistent with analyticity, perturbative unitarity, cluster decomposition, and the assumed symmetry principles.

In summary, the EFT program consists of the following steps:

- Identify the soft and hard scales, and the degrees of freedom appropriate for (low-energy) nuclear physics.
- Identify the relevant symmetries of low-energy QCD and investigate if and how they are broken.
- Construct the most general Lagrangian consistent with those symmetries and symmetry breakings.
- Design an organizational scheme that can distinguish between more and less important contributions: a low-momentum expansion.
- Guided by the expansion, calculate Feynman diagrams for the problem under consideration to the desired accuracy.

Note that we are presenting here the *chiral* EFT which scales with the ρ-mass and includes pion degrees of freedom. However, the ρ-mass is not the only possible (hard) scale. Depending on what energies we are interested in, other scales may be more appropriate. For example, if we wish to focus on a nuclear scenario at very low energy ($< m\pi$), then it is suggestive to choose the pion-mass, $m\pi$, as scale. Such EFT can, of course, not have pion degrees of freedom anymore and is, therefore, known as pionless EFT. It consists of contact terms, only. Thus, this pionless EFT may appear forbiddingly simplistic at first glance, but, as it turns out, it has some surprisingly intriguing features, particularly, with regard to renormalization and power counting. Since it is beyond the scope of this article to discuss the pionless EFT in detail, we like to refer the interested reader to the excellent review articles.

Symmetries of Low-energy QCD and their Breakings

The QCD Lagrangian reads

$$\mathcal{L}_{QCD} = \bar{q}(i\gamma^{\mu} D_{\mu} - M)q - \frac{1}{4} G_{\mu\nu,a} G_{a}^{\mu\nu}$$

with the gauge-covariant derivative

$$D_\mu = \partial_\mu - ig\frac{\lambda_a}{2}A_{\mu,a}$$

and the gluon field strength tensor (note that for *SU*(*N*) group indices, we use Latin letters, ...,*a*,*b*,*c*,...,*i*,*j*,*k*,..., and, in general, do not distinguish between subscripts and superscripts)

$$G_{\mu\nu,a} = \partial_\mu A_{\nu,a} - \partial_\nu A_{\mu,a} + gf_{abc}A_{\mu,b}A_{\nu,c}.$$

In the above, q denotes the quark fields and M the quark mass matrix. Further, g is the strong coupling constant and $A_{\mu,a}$ are the gluon fields. The λ_a are the Gell-Mann matrices and the *fabc* the structure constants of the $SU(3)_{color} \,|\, Lie\ algebra\ (a,b,c=1,\ldots,8)$; summation over repeated indices is always implied. The gluon-gluon term in the last equation arises from the non-Abelian nature of the gauge theory and is the reason for the peculiar features of the color force.

The masses of the up (u), down (d), and strange (s) quarks are (PDG):

$$m_u = 2.3 \pm 0.7\,\text{MeV},$$
$$m_d = 4.8 \pm 0.7\,\text{MeV},$$
$$m_s = 95 \pm 5\,\text{MeV}.$$

These masses are small as compared to a typical hadronic scale, i.e., a scale of low-mass hadrons which are not Goldstone bosons, e.g., m_ρ=0.78 GeV≈1 GeV.

It is therefore of interest to discuss the QCD Lagrangian in the limit of vanishing quark masses:

$$\mathcal{L}^0_{QCD} = \bar{q}i\gamma^\mu D_\mu q - \frac{1}{4}G_{\mu\nu,a}G^{\mu\nu}_a.$$

Defining right- and left-handed quark fields,

$$q_R = P_R q,\ \ q_L = P_L q,$$

With

$$P_R = \frac{1}{2}(1+\gamma_5),\ \ P_L\frac{1}{2}(1-\gamma_5)$$

we can rewrite the Lagrangian as follows:

$$\mathcal{L}^0_{QCD} = \bar{q}_R{}^i\gamma^\mu D_\mu q_R + \bar{q}L^i\gamma^\mu D_\mu q_L - \frac{1}{4}G_{\mu\nu,a}G^{\mu\nu}_a.$$

This equation reveals that *the right- and left-handed components of massless quarks do not mix* in the QCD Lagrangian. For the two-flavor case, this is $SU(2)_R \times SU(2)_L$ symmetry, also known as *chiral symmetry*. However, this symmetry is broken in two ways: explicitly and spontaneously.

Explicit Symmetry Breaking

The mass term $-\bar{q}M_q$ in the QCD Lagrangian Equation $\mathcal{L}^0_{QCD} = \bar{q}i\gamma^\mu D_\mu q - \frac{1}{4}G_{\mu\nu,a}G^{\mu\nu}_a$. breaks chiral symmetry explicitly. To better see this, let's rewrite M for the two-flavor case,

$$\begin{aligned}\mathcal{M} &= \begin{pmatrix} m_u & 0 \\ 0 & m_d \end{pmatrix} \\ &= \frac{1}{2}(m_u + m_d)\begin{pmatrix} 1 & 0 \\ 0 & 1 \end{pmatrix} + \frac{1}{2}(m_u - m_d)\begin{pmatrix} 1 & 0 \\ 0 & -1 \end{pmatrix} \\ &= \frac{1}{2}(m_u + m_d)I + \frac{1}{2}(m_u - m_d)\tau_3.\end{aligned}$$

The first term in the last equation in invariant under $SU(2)_V$ (isospin symmetry) and the second term vanishes for $m_u = m_d$. Thus, isospin is an exact symmetry if $m_u = m_d$. However, both terms in above Equation break chiral symmetry. Since the up and down quark masses [eqs. $m_u = 2.3 \pm 0.7\,\text{MeV}$, and $m_d = 4.8 \pm 0.7\,\text{MeV}$,] are small as compared to the typical hadronic mass scale of ~1 GeV, the explicit chiral symmetry breaking due to non-vanishing quark masses is very small.

Spontaneous Symmetry Breaking

A (continuous) symmetry is said to be *spontaneously broken* if a symmetry of the Lagrangian is not realized in the ground state of the system. There is evidence that the (approximate) chiral symmetry of the QCD Lagrangian is spontaneously broken for dynamical reasons of nonperturbative origin which are not fully understood at this time. The most plausible evidence comes from the hadron spectrum.

From chiral symmetry, one naively expects the existence of degenerate hadron multiplets of opposite parity, i.e., for any hadron of positive parity one would expect a degenerate hadron state of negative parity and vice versa. However, these 'parity doublets' are not observed in nature. For example, take the ρ-meson which is a vector meson of negative parity ($J^P = 1^-$) and mass 776 MeV. There does exist a 1^+meson, the a_1, but it has a mass of 1230 MeV and, therefore, cannot be perceived as degenerate with the ρ. On the other hand, the ρ meson comes in three charge states (equivalent to three isospin states), the $\rho^\pm$ and the ρ^0, with masses that differ by at most a few MeV. Thus, in the hadron spectrum, $SU(2)_V$ (isospin) symmetry is well observed, while axial symmetry is broken: $SU(2)_R \times SU(2)_L$ is broken down to $SU(2)_V$.

A spontaneously broken global symmetry implies the existence of (massless) Goldstone bosons. The Goldstone bosons are identified with the isospin triplet of the (pseudoscalar) pions, which explains why pions are so light. The pion masses are not exactly zero because the up and down quark masses are not exactly zero either (explicit symmetry breaking). Thus, pions are a truly remarkable species: they reflect spontaneous as well as explicit symmetry breaking. Goldstone bosons interact weakly at low energy. They are degenerate with the vacuum and, therefore, interactions between them must vanish at zero momentum and in the chiral limit ($m_\pi \to 0$).

Chiral Effective Lagrangians

The next step in the EFT program is to build the most general Lagrangian consistent with the (broken) symmetries. An elegant formalism for the construction of such Lagrangians was developed by Callan , who worked out the group-theoretical foundations of non-linear realizations of chiral symmetry. It is characteristic for these non-linear realizations that, whenever functions of the Goldstone bosons appear in the Lagrangian, they are always accompanied with at least one space-time derivative. The Lagrangians given below are built upon the Callan formalism.

As discussed, the relevant degrees of freedom are pions (Goldstone bosons) and nucleons. Since the interactions of Goldstone bosons must vanish at zero momentum transfer and in the chiral limit ($m_\pi \to 0$), the low-energy expansion of the Lagrangian is arranged in powers of derivatives and pion masses. The hard scale is the chiral-symmetry breaking scale, $\Lambda\chi \approx 1$ GeV. Thus, the expansion is in terms of powers of $Q / \Lambda\chi$ where Q is a (small) momentum or pion mass. This is the essence of chiral perturbation theory (ChPT).

The effective Lagrangian can formally be written as,

$$\mathcal{L}_{eff} = \mathcal{L}_{\pi\pi} + \mathcal{L}_{\pi N} + \mathcal{L}_{NN} + \ldots,$$

where $\mathcal{L}_{\pi\pi}$ deals with the dynamics among pions, $\mathcal{L}_{\pi N}$ describes the interaction between pions and a nucleon, and L_{NN} contains two-nucleon contact interactions which consist of four nucleon-fields (four nucleon legs) and no meson fields. The ellipsis stands for terms that involve two nucleons plus pions and three or more nucleons with or without pions, relevant for nuclear many-body forces (cf. last two terms of equation below). The individual Lagrangians are organized as follows:

$$\mathcal{L}_{\pi\pi} = \mathcal{L}_{\pi\pi}^{(2)} + \mathcal{L}_{\pi\pi}^{(4)} + \ldots,$$

$$\mathcal{L}_{\pi N} = \mathcal{L}_{\pi N}^{(1)} + \mathcal{L}_{\pi N}^{(2)} + \mathcal{L}_{\pi N}^{(3)} + \ldots,$$

and

$$\mathcal{L}_{NN} = \mathcal{L}_{NN}^{(0)} + \mathcal{L}_{NN}^{(2)} + \mathcal{L}_{NN}^{(4)} + \ldots,$$

where the superscript refers to the number of derivatives or pion mass insertions (chiral dimension) and the ellipsis stands for terms of higher dimensions.

Above, we have organized the Lagrangians by the number of derivatives or pion-mass insertions. This is the standard way, appropriate particularly for considerations of π-π and π-N scattering. As it turns out, for interactions among nucleons, it is sometimes more useful to consider the so-called index of the interaction,

$$\Delta \equiv d + \frac{n}{2} - 2,$$

where d is the number of derivatives or pion-mass insertions and n the number of nucleon field operators (nucleon legs). We will now write down the Lagrangian in terms of increasing values of the parameter Δ and we will do so using the so-called Heavy-Baryon (HB) formalism which we indicate by a "hat".

The lowest-index Lagrangian reads,

$$\begin{aligned}
\hat{\mathcal{L}}^{\Delta=0} &= \frac{1}{2}\partial_u\pi.\partial^u\pi - \frac{1}{2}m_\pi^2\pi^2 \\
&+ \frac{1-4\alpha}{2f_\pi^2}(\pi.\partial_u\pi)(\pi.\partial^u\pi) - \frac{\alpha}{f_\pi^2}\pi^2\partial_u\pi.\partial^u\pi + \frac{8\alpha-1}{8f_\pi^2}m_\pi^2\pi^4 \\
&+ \bar{N}\left[i\partial_0 - \frac{g_A}{2f_\pi}\tau.(\vec{\sigma}.\vec{\nabla})\pi - \frac{1}{4f_\pi^2}\tau.(\pi\times\partial_0\pi)\right]N \\
&+ \bar{N}\left\{\frac{g_A(4\alpha-1)}{4f_\pi^3}(\tau.\pi)\left[\pi.(\vec{\sigma}.\vec{\nabla})\pi\right] + \frac{g_A\alpha}{2f_\pi^3}\pi^2\left[\tau.(\vec{\sigma}.\vec{\nabla})\pi\right]\right\}N \\
&- \frac{1}{2}C_S\bar{N}N\bar{N}N - \frac{1}{2}C_T(\bar{N}\vec{\sigma}N).(\bar{N}\vec{\sigma}N) + \ldots,
\end{aligned}$$

and higher-index Lagrangians are,

$$\begin{aligned}
\hat{\mathcal{L}}^{\Delta=1} &= \bar{N}\left\{\frac{\vec{\nabla}^2}{2M_N} - \frac{ig_A}{4M_Nf_\pi}\tau.\left[\vec{\sigma}.(\overleftarrow{\nabla}\partial_0\pi - \partial_0\pi\vec{\nabla})\right]\right. \\
&\left. - \frac{i}{8M_Nf_\pi^2}\tau.\left[\overleftarrow{\nabla}.(\pi\times\vec{\nabla}\pi) - (\pi\times\vec{\nabla}\pi).\vec{\nabla}\right]\right\}N \\
&+ \bar{N}\left[4c_1m_\pi^2 - \frac{2c1}{f_\pi^2}m_\pi^2\pi^2 + \left(c_2 - \frac{g_A^2}{8M_N}\right)\frac{1}{f_\pi^2}(\partial_0\pi.\partial_0\pi)\right. \\
&\left. + \frac{c_3}{f_\pi^2}(\partial_u\pi.\partial^u\pi) - \left(c_4 + \frac{1}{4M_N}\right)\frac{1}{f_\pi^2}\in^{ijk}\in^{abc}\sigma^i\tau^a(\partial^j\pi^b)(\partial^k\pi^c)\right]N \\
&- \frac{D}{4f_\pi}(\bar{N}N)\bar{N}\left[\tau.(\vec{\sigma}.\vec{\nabla})\pi\right]N - \frac{1}{2}E(\bar{N}N)(\bar{N}\tau N).(\bar{N}\tau N) + \ldots, \\
\hat{\mathcal{L}}^{\Delta=2} &= \mathcal{L}_{\pi\pi}^{(4)} + \mathcal{L}_{\pi N}^{(3)} + \hat{\mathcal{L}}_{NN}^{(2)} + \ldots, \\
\hat{\mathcal{L}}^{\Delta=4} &= \hat{\mathcal{L}}_{NN}^{(4)} + \ldots,
\end{aligned}$$

where the ellipses represent terms that are irrelevant for the derivation of nuclear forces up to fourth order. The Lagrangian $\hat{\mathcal{L}}_{\pi N}^{(3)}$ can be found in and the *NN* contact Lagrangians are given below. The pion fields are denoted by π and the heavy baryon nucleon field by N ($\bar{N}=N\dagger$). Furthermore, *gA*, *fπ*, *mπ*, and *MN* are the axial-vector coupling constant, pion decay constant, pion mass, and nucleon mass, respectively. Numerical values for these quantities will be given later. The *ci* are Low-Energy Constants (LECs) from the dimension two πN Lagrangian and α is a parameter that appears in the expansion of the pion fields. Results are independent of α. The πNN coupling constant, $f_{\pi NN}$, used in the derivation of the one-pion-exchange potential in figure, is related to the above quantities by

$$\frac{f_{\pi NN}}{m_\pi}=\frac{g_A}{2f_\pi},$$

cf. Equation $V_{1\pi}(\vec{p}',\vec{p})=-\frac{g_A^2}{4f_\pi^2}T_1\bullet T_2\frac{\vec{\sigma}_1\bullet\vec{q}\,\vec{\sigma}_2\bullet\vec{q}}{q^2+m_\pi^2}$ below.

The lowest order (or leading order) *NN* Lagrangian has no derivatives and reads

$$\hat{\mathcal{L}}_{NN}^{(0)}=-\frac{1}{2}C_S\bar{N}N\bar{N}N-\frac{1}{2}C_T\left(\bar{N}\vec{\sigma}N\right).\left(\bar{N}\vec{\sigma}N\right),$$

where C_S and C_T are unknown constants which are determined by a fit to the *NN* data. The second order *NN* Lagrangian can be stated as follows,

$$\begin{aligned}\hat{\mathcal{L}}_{NN}^{(2)}=&-C_1'\left[\left(\bar{N}\vec{\nabla}N\right)^2+\left(\overline{\vec{\nabla}N}N\right)^2\right]-C_2'\left(\bar{N}\vec{\nabla}N\right).\left(\overline{\vec{\nabla}N}N\right)-C_3'\bar{N}N\left[\bar{N}N\right]\left[\bar{N}\vec{\nabla}^2N+\overline{\vec{\nabla}^2N}N\right]\\&-iC_4'\left[\bar{N}\vec{\nabla}N.\left(\overline{\vec{\nabla}N}\times\vec{\sigma}N\right)+\left(\overline{\vec{\nabla}N}\right)N.\left(\bar{N}\vec{\sigma}\times\vec{\nabla}N\right)\right]\\&-iC_5'\bar{N}N\left(\overline{\vec{\nabla}N}.\vec{\sigma}\times\vec{\nabla}N\right)-iC_6'\left(\bar{N}\vec{\sigma}N\right).\left(\overline{\vec{\nabla}N}\times\vec{\nabla}N\right)\\&-(C_7'\delta_{ik}\delta_{jl}+C_8'\delta_{il}\delta_{kj}+C_9'\delta_{ij}\delta_{kl})[\bar{N}\sigma_k\partial_i N\bar{N}\sigma_l\partial_j N+\overline{\partial_i N}\sigma_k N\overline{\partial_j N}\sigma_l N]\\&-(C_{10}'\delta_{ik}\delta_{jl}+C_{11}'\delta_{il}\delta_{kj}+C_{12}'\delta_{ij}\delta_{kl})\bar{N}\sigma_k\partial_i N\overline{\partial_j N}\sigma_l N\\&-\left(\frac{1}{2}C_{13}'\left(\delta_{ik}\delta_{jl}+\delta_{il}\delta_{kj}\right)+C_{14}'\delta_{ij}\delta_{kl}\right)\left[\overline{\partial_i N}\sigma_k\partial_j N+\overline{\partial_j N}\sigma_k\partial_i N\right]\bar{N}\sigma_l N\end{aligned}$$

Similar to C_S and C_T of above Equation, the C'_i of the above equation are unknown constants which are fixed by a fit to the *NN* data. Obviously, these contact Lagrangians blow up quite a bit with increasing order, which is why we do not give L^(4)*NN* explicitly here. The *NN* contact potentials that emerge from these Lagrangians are given below.

Power Counting

Effective Lagrangians have infinitely many terms, and an unlimited number of Feynman graphs can be calculated from them. Therefore, we need a scheme that makes the theory manageable and calculable. This scheme which tells us how to distinguish between large (important) and small (unimportant) contributions is chiral perturbation theory (ChPT).

In ChPT, graphs are analyzed in terms of powers of small external momenta over the large scale: $(Q/\Lambda\chi)^v$, where *Q* is generic for a momentum (nucleon three-momentum or pion four-momentum) or a pion mass and $\Lambda_\chi \sim 1 GeV$ is the chiral symmetry breaking scale (hadronic scale, hard scale). Determining the power *v* has become known as power counting.

The nuclear potential is assembled from irreducible graphs. By definition, an irreducible graph is a diagram that cannot be separated into two by cutting only nucleon lines. Following the Feynman rules of covariant perturbation theory, a nucleon propagator is Q^{-1}, a pion propagator Q^{-2}, each derivative in any interaction is *Q*, and each four-momentum integration Q^4. This is also known as naive dimensional analysis. Applying then some topological identities, one obtains for the power of an irreducible diagram involving A nucleons

$$\nu = -2 + 2A - 2C + 2L + \sum_i \Delta_i,$$

with

$$\Delta_i \equiv d_i + \frac{n_i}{2} - 2,$$

where C denotes the number of separately connected pieces and L the number of loops in the diagram; d_i is the number of derivatives or pion-mass insertions and n_i the number of nucleon fields (nucleon legs) involved in vertex i; the sum runs over all vertices i contained in the diagram under consideration. Note that $\Delta_i \geq 0$ for all interactions allowed by chiral symmetry. Purely pionic interactions have at least two derivatives ($d_i \geq 2, n_i = 0$); interactions of pions with a nucleon have at least one derivative ($d_i \geq 1, n_i = 2$); and nucleon-nucleon contact terms ($n_i = 4$) have $d_i \geq 0$. This demonstrates how chiral symmetry guarantees a low-energy expansion.

The power formula equation $\nu = -2 + 2A - 2C + 2L + \sum_i \Delta_i$, allows to predict the leading orders of connected multi-nucleon forces. Consider a m-nucleon irreducibly connected diagram (m-nucleon force) in an A-nucleon system ($m \leq A$). The number of separately connected pieces is $C=A-m+1$. Inserting this into Equation $\nu = -2 + 2A - 2C + 2L + \sum_i \Delta_i$, together with $L=0$ and $\sum_i \Delta_i = 0$, yields $\nu=2m-4$. Thus, two-nucleon forces ($m=2$) start at $\nu=0$, three-nucleon forces ($m=3$) at $\nu=2$ (but they happen to cancel at that order), and four-nucleon forces at $\nu=4$ (they don't cancel).

For later purposes, we note that for an irreducible *NN* diagram ($A=2$, $C=1$), the power formula collapses to the very simple expression

$$\nu = 2L + \sum_i \Delta_i$$

In summary, the chief point of the ChPT expansion is that, at a given order ν, there exists only a finite number of graphs. This is what makes the theory calculable. The expression $(Q/\Lambda_\chi)^{\nu+1}$ provides a rough estimate of the relative size of the contributions left out and, thus, of the accuracy at order ν. In this sense, the theory can be calculated to any desired accuracy and has predictive power.

Hierarchy of Nuclear Forces: Overview

Chiral perturbation theory and power counting imply that nuclear forces emerge as a hierarchy controlled by the power ν, above in figure.

In lowest order, better known as leading order (LO, ν=0), the *NN* potential is made up by two momentum-independent contact terms ($\sim Q^0$), represented by the four-nucleon-leg graph with a small dot shown in the first row of figure. Furthermore, there is the static one-pion exchange (1PE), second diagram in the first row of the figure. This is, of course, a rather rough approximation to the two-nucleon force (2NF), but accounts already for some important features. The 1PE provides the tensor force, necessary to describe the deuteron, and it explains *NN* scattering in peripheral partial waves of very high orbital angular momentum. At this order, the two contacts, which contribute only in *S*-waves, provide the short- and intermediate-range interaction, which is somewhat crude.

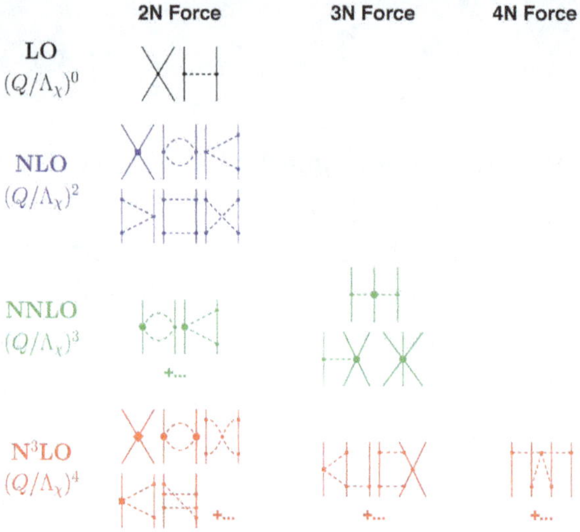

Figure: Hierarchy of nuclear forces in ChPT. Solid lines represent nucleons and dashed lines pions. Small dots, large solid dots, solid squares, and solid diamonds denote vertices of index $\Delta_i = 0, 1, 2$, and 4, respectively.

In the next order, ν=1, all contributions vanish due to parity and time-reversal invariance.

Therefore, the next-to-leading order (NLO) is ν=2. Two-pion exchange (2PE) occurs for the first time ("leading 2PE") and, thus, the creation of a more sophisticated description of the intermediate-range interaction is starting here. Since the loop involved in each 2PE-diagram implies already ν=2 ; $\nu = 2L + \sum_i \Delta_i$, the vertices must have Δ_i=0. Therefore, at this order, only the lowest order πNN and $\pi\pi NN$ vertices are allowed which is why the leading 2PE is rather weak. Furthermore, there are seven new contact terms of $O(Q^2)$, shown in the figure by the four-nucleon-leg graph with a solid square, which contribute in *S* and *P* waves. The operator structure of these contacts include a spin-orbit term besides central, spin-spin, and tensor terms. Thus, essentially all spin-isospin structures necessary to describe the two-nucleon force phenomenologically have been generated. The main deficiency at this stage of development is an insufficient intermediate-range attraction.

This problem is finally fixed at order three (ν=3), next-to-next-to-leading order (NNLO). The 2PE involves now the two-derivative $\pi\pi NN$ seagull vertices (proportional to the *ci* LECs) denoted by a large solid dot in above Figure. These vertices represent correlated 2PE as well as intermediate Δ (1232)-isobar contributions. It is well-known from the meson phenomenology of nuclear forces that these two contributions are crucial for a realistic and quantitative 2PE model. Consequently, the 2PE now assumes a realistic size and describes the intermediate-range attraction of the nuclear force about right . Moreover, first relativistic corrections come into play at this order. There are no new contacts, because contacts appear only at even orders.

The reason why we talk of a hierarchy of nuclear forces is that two- and many-nucleon forces are created on an equal footing and emerge in increasing number as we go to higher and higher orders. At NNLO, the first set of nonvanishing three-nucleon forces (3NFs) occur, cf. column '3N Force' of in above figure. In fact, at the previous order, NLO, irreducible 3N graphs appear already, however, it can be shown that these diagrams all cancel. Since nonvanishing 3NF contributions happen first at order $(Q/\Lambda\chi)_3$, they are very weak as compared to the 2NF which starts at $(Q/\Lambda_\chi)^0$.

More 2PE is produced at *v*=4, next-to-next-to-next-to-leading order (N3LO), of which we show only a few symbolic diagrams in above figure. Two-loop 2PE graphs show up for the first time and so does three-pion exchange (3PE) which necessarily involves two loops. The 3PE is negligible at this order. Most importantly, 15 new contact terms ~*Q*4 arise and are represented by the four-nucleon-leg graph with a solid diamond. They include a quadratic spin-orbit term and contribute up to *D*-waves.

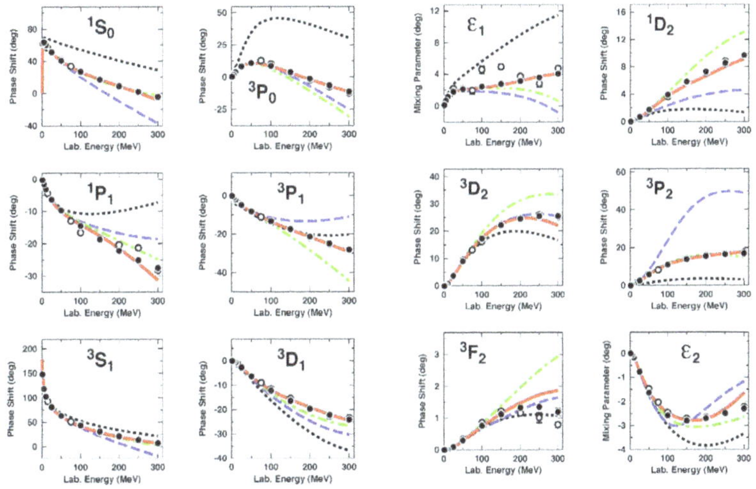

Figure: Phase shifts of *np* scattering as calculated from *NN* potentials at different orders of ChPT. The black dotted line is LO, the blue dashed is NLO, the green dash-dotted NNLO, and the red solid N3LO. Partial waves with total angular momentum *J*≤2 are displayed. The solid dots and open circles are the results from phase shift analysis.

Mainly due to the larger number of contact terms, a quantitative description of the two-nucleon interaction up to about 300 MeV lab. energy is possible, at N3LO. This is demonstrated in above Figure, where we show the order by order improvement of the *NN* phase shift predictions from LO to N3LO. Table quantifies this order by order improvement by providing the χ^2 /datum for the fit of *NN* data from NLO to N^3LO.

*T*lab **bin (MeV)**	**number of *np* data**	**N3LO**	**NNLO**	**NLO**
0--100	1058	1.05	1.7	4.5
100--190	501	1.08	22	100
190--290	843	1.15	47	180
0--290	**2402**	**1.10**	**20**	**86**

Table: χ2/datum for the reproduction of the np data below 290 MeV by chiral NN potentials at NLO, NNLO, and N3LO. T_{lab} *denotes the kinetic energy of the incident nucleon in the laboratory system.*

Moreover, there are more 3NF contributions at N^3LO, and four-nucleon forces (4NFs) start at this order. Since the leading 4NFs come into existence one order higher than the leading 3NFs, 4NFs are weaker than 3NFs. Thus, ChPT provides a straightforward explanation for the empirically known fact that 2NF ≫ 3NF ≫ 4NF.

Comparison with Conventional Meson Theory

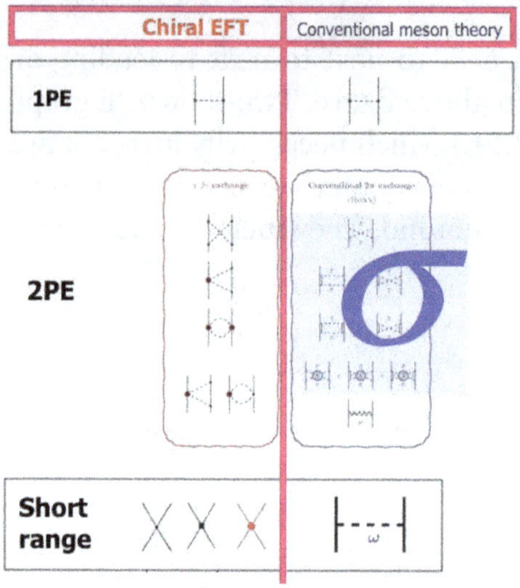

Figure: Comparing the chiral EFT approach to the *NN*interaction (left) with conventional meson theory (right).

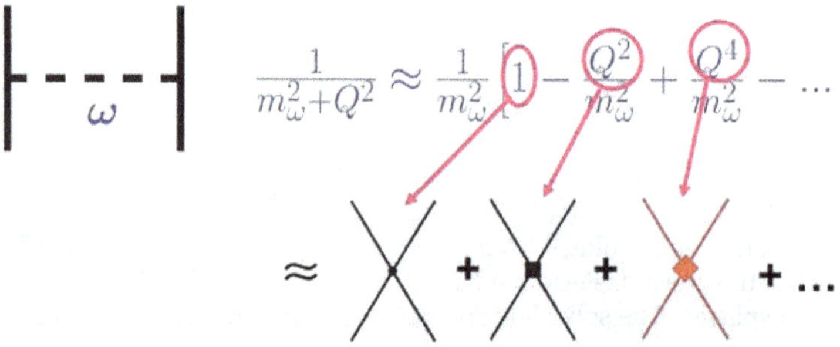

Figure: Relationship between heavy boson exchange and the contact terms of chiral EFT.

We have now two approaches at hand that can both describe the *NN* interaction quantitatively, which is a non-trivial result. It is then natural to ask, in which way the two approaches differ. There is a clear and revealing answer.

In chiral EFT, the nuclear potential is expanded in terms of increasing powers of small momenta, $(Q / \Lambda\chi)^{v}$. In meson theory, the expansion is in terms of Yukawas, $Y(m\alpha r)$, of increasing masses $m\alpha$, corresponding to decreasing ranges $1 / m_{\alpha}$.

Since both approaches describe the same complicated object quantitatively, they should be equivalent to a large extent. This is demonstrated in above figure. First, there is a 1PE in both cases, which is trivial. The 2PE may look diagrammatically quite different, but the figure shows the correspondence between the contributions. The main difference is that, in chiral EFT, the 2PE is build up order by order, while in conventional meson theory it comes as one set. Finally, the short-range contributions appear graphically very different with heavy boson exchange (like ω-exchange) in meson theory and contacts in chiral EFT. However, since $Q \ll m\omega \approx \Lambda\chi$, the propagator of a heavy-meson can be expanded into a power series generating contacts of increasing order, as demonstrated in above figure.

Although the two approaches can be regarded as equivalent, there are arguments why chiral EFT may be perceived as superior. Chiral EFT

- Is more closely connected to QCD via chiral symmetry;
- Comes with an organizational scheme (power counting) that allows to estimate the accuracy of the predictions (at a given order);
- Generates two- and many-body forces on an equal footing.

Two-nucleon Forces: Doing the Math

Here we will fill in the mathematical details we left out when presenting the overview over the chiral hierarchy. Up to N^3LO, the various irreducible diagrams, which were shown in above figure and define the chiral *NN* potential order by order, are given by:

$$V_{LO} = V_{ct}^{(0)} + V_{1\pi}^{(0)}$$
$$V_{NLO} = V_{LO} + V_{ct}^{(2)} + V_{1\pi}^{(2)} + V_{2\pi}^{(2)}$$
$$V_{NNLO} = V_{NLO} + V_{1\pi}^{(3)} + V_{2\pi}^{(3)}$$
$$V_{N^3LO} = V_{NNLO} + V_{ct}^{(4)} + V_{1\pi}^{(4)} + V_{2\pi}^{(4)} + V_{3\pi}^{(4)}$$

where the superscript denotes the order v of the low-momentum expansion. Contact potentials carry the subscript "ct" and pion-exchange potentials can be identified by an obvious subscript.

The charge-independent 1PE potential reads

$$V_{1\pi}(\vec{p}\,',\vec{p}) = -\frac{g_A^2}{4f_\pi^2}\tau_1 \cdot \tau_2 \frac{\vec{\sigma}_1 \cdot \vec{q}\,\vec{\sigma}_2 \cdot \vec{q}}{q^2 + m_\pi^2}$$

Numerical vales for the constants are fπ=92.4 MeV, gA=1.29; and mπ=138 MeV for the average pion mass (above figure and Equation $\frac{f_{\pi NN}}{m_\pi} = \frac{g_A}{2f_\pi}$,). Since higher order corrections contribute only to mass and coupling constant renormalizations and since, on shell, there are no relativistic corrections, the on-shell 1PE has the same form as in Equation $V_{1\pi}(\vec{p}\,',\vec{p}) = -\frac{g_A^2}{4f_\pi^2}\tau_1 \cdot \tau_2 \frac{\vec{\sigma}_1 \cdot \vec{q}\,\vec{\sigma}_2 \cdot \vec{q}}{q^2 + m_\pi^2}$ at all orders.

The two zero-order contact terms at LO are

$$V_{ct}^{(0)}(\vec{p}\,',\vec{p}) = C_s + C_T \vec{\sigma}_1 \cdot \vec{\sigma}_2$$

To state the mathematical expressions for 2PE contributions, we use the following scheme:

$$\begin{aligned} V_{2\pi}^{(v)}(\vec{p}',\vec{p}) = {} & V_C^{(v)} + \tau_1 \cdot \tau_2 W_C^{(v)} \\ & + \left[V_S^{(v)} + \tau_1 \cdot \tau_2 W_S^{(v)}\right] \vec{\sigma}_1 \cdot \vec{\sigma}_2 \\ & + \left[V_{LS}^{(v)} + \tau_1 \cdot \tau_2 W_{LS}^{(v)}\right]\left(-i\vec{S} \cdot (\vec{q} \times \vec{k})\right) \\ & + \left[V_T^{(v)} + \tau_1 \cdot \tau_2 W_T^{(v)}\right] \vec{\sigma}_1 \cdot \vec{q} \vec{\sigma}_2 \cdot \vec{q} \\ & + \left[V_{\sigma L}^{(v)} + \tau_1 \cdot \tau_2 W_{\sigma L}^{(v)}\right] \vec{\sigma}_1 \cdot (\vec{q} \times \vec{k}) \vec{\sigma}_2 \cdot (\vec{q} \times \vec{k}), \end{aligned}$$

where $\vec{p}'$ and $\vec{p}$ denote the final and initial nucleon momenta in the CMS, respectively; moreover,

$\vec{q} \equiv \vec{p}' - \vec{p}$ is the momentum transfer,

$\vec{k} \equiv \frac{1}{2}(\vec{p}' + \vec{p})$ the average momentum,

$\vec{S} \equiv \frac{1}{2}(\vec{\sigma}_1 + \vec{\sigma}_2)$ the total spin,

And $\vec{\sigma}_{1,2}$ and $\tau_{1,2}$ are the spin and isospin operators, respectively, of nucleon 1 and 2. For on-energy-shell scattering, V_i and W_i ($i=C,S,LS,T,\sigma L$) can be expressed as functions of q and k, only (with $q \equiv |\vec{q}|$ and $k \equiv |\vec{k}|$).

Using the above scheme, the contribution from the five NLO 2PE diagrams can be stated in an amazingly compact form, namely,

$$W_e^{(2)} = -\frac{L(q)}{\text{ü} \quad \pi^{\text{ü}} f_\pi}\left[4m_\pi^2\left(5g_A^4 - 4g_A^2 - 1\right) + q^2\left(23g_A^4 - 10g_A^2 - 1\right) + \frac{48g_A^4 m_\pi^4}{w}\right],$$

$$V_T^{(2)} \aleph \frac{1}{q^{\text{ü}}} V_S^{(2)} \quad \frac{3g_A^4 m_\pi^4}{64\pi \; f_\pi},$$

Where

$$L(q) \equiv \frac{w}{q} \text{In} \frac{w+q}{2m_\pi}$$

And

$$w \equiv \sqrt{4m_\pi^2 + q^2}.$$

Note that all 2π loops involve a four-dimensional integral, which is divergent. Thus regularization is required. All 2π contributions given in this subsection are obtained by applying dimensional regularization (DR).

The seven NLO contact terms are:

$$\begin{aligned} V_{ct}^{(0)}(\vec{p}',\vec{p}) = {} & C_1 q^2 + C_2 k^2 \\ & + (C_3 q^2 + C_4 k^2)\vec{\sigma}_1.\vec{\sigma}_2 \\ & + C_5(-i\vec{S}\cdot(\vec{q}\times\vec{k})) \\ & + C_6(\vec{\sigma}_1\cdot\vec{q}\,)(\vec{\sigma}_2\cdot\vec{q}) \\ & + C_7(\vec{\sigma}_1.\vec{k}\,)(\vec{\sigma}_2.\vec{k}\,). \end{aligned}$$

The coefficients C_i used here in the contact potential are, of course, related to the coefficients C'_i that occur in the Lagrangian $\hat{\mathcal{L}}_{NN}^{(2)}$ Equation

$$\begin{aligned} \hat{\mathcal{L}}_{NN}^{(2)} = {} & -C'_1\left[\left(\bar{N}\vec{\nabla}N\right)^2 + \left(\overline{\vec{\nabla}N}N\right)^2\right] - -C'_2\left(\bar{N}\vec{\nabla}N\right).\left(\overline{\vec{\nabla}N}N\right) - C'_3\bar{N}N\left[\bar{N}N\right]\left[\bar{N}\vec{\nabla}^2N + \overline{\vec{\nabla}^2N}N\right] \\ & -iC'_4\left[\bar{N}\vec{\nabla}N.\left(\overline{\vec{\nabla}N}\times\vec{\sigma}N\right) + \left(\overline{\vec{\nabla}N}\right)N.\left(\bar{N}\vec{\sigma}\times\vec{\nabla}N\right)\right] \\ & -iC'_5\bar{N}N\left(\overline{\vec{\nabla}N}.\vec{\sigma}\times\vec{\nabla}N\right) - iC'_6\left(\bar{N}\vec{\sigma}N\right).\left(\overline{\vec{\nabla}N}\times\vec{\nabla}N\right) \\ & -(C'_7\delta_{ik}\delta_{jl} + C'_8\delta_{il}\delta_{kj} + C'_9\delta_{ij}\delta_{kl})[\bar{N}\sigma_k\partial_i N\bar{N}\sigma_l\partial_j N + \overline{\partial_i N}\sigma_k N\overline{\partial_j N}\sigma_l N] \\ & -(C'_{10}\delta_{ik}\delta_{jl} + C'_{11}\delta_{il}\delta_{kj} + C'_{12}\delta_{ij}\delta_{kl})\bar{N}\sigma_k\partial_i N\overline{\partial_j N}\sigma_l N \\ & -\left(\frac{1}{2}C'_{13}\left(\delta_{ik}\delta_{jl} + \delta_{il}\delta_{kj}\right) + C'_{14}\delta_{ij}\delta_{kl}\right)\left[\overline{\partial_i N}\sigma_k\partial_i N + \overline{\partial_j N}\sigma_k\partial_i N\right]\bar{N}\sigma_l N \end{aligned},$$

but, the exact relationship is unimportant.

The NNLO 2PE is represented by the following expressions:

$$V_C^{(3)} = \frac{3g_A^2}{16\pi f_\pi^4}\left\{\frac{g_A^2 m_\pi^5}{16M_N w^2} - \left[2m_\pi^2\left(2c_1 - c_3\right) - q^2\left(c_3 + \frac{3g_A^2}{16M_N}\right)\right]\tilde{w}^2 A(q)\right\},$$

$$W_C^{(3)} = \frac{g_A^2}{128\pi M_N f_\pi^4}\left\{3g_A^2 m_\pi^5 w^{-2} - \left[4m_\pi^2 + 2q^2 - g_A^2\left(4m_\pi^2 + 3q^2\right)\right]\tilde{w}^2 A(q)\right\},$$

$$V_T^{(3)} = \frac{1}{q^2}V_S^{(3)} = \frac{9g_A^2\tilde{w}^2 A(q)}{512\pi M_N f_\pi^4},$$

$$W_T^{(3)} = -\frac{1}{q^2}W_S^{(3)} = \frac{g_A^2 A(q)}{32\pi f_\pi^4}\left[\left(c_4 + \frac{1}{M_N}\right)w^2 - \frac{g_A^2}{8M_N}\left(10m_\pi^2 + 3q^2\right)\right],$$

$$V_{LS}^{(3)} = \frac{3g_A^2\tilde{w}^2 A(q)}{32\pi M_N f_\pi^4},$$

$$W_{LS}^{(3)} = \frac{g_A^2\left(1 - g_A^2\right)}{32\pi M_N f_\pi^4}w^2 A(q),$$

with

$$A(q) \equiv \frac{1}{2q}\arctan\frac{q}{2m_\pi}$$

And

$$\tilde{w} \equiv \sqrt{2m_\pi^2 + q^2}.$$

This contribution to the 2PE is the crucial one, because it provides an intermediate-range attraction of proper strength. The iso-scalar central potential, $V_C^{(3)}$, is strong and attractive due to the LEC c3, which is negative and of large magnitude. Via resonance saturation, c3 is associated with π-π correlations ('σ meson') and virtual Δ-isobar excitations, which create the most crucial contributions to 2PE in the frame work of conventional meson theory. First relativistic $1/MN$ corrections come also into play at this order; they are included in the above potential expressions.

The contacts at N^3LO are:

$$\begin{aligned} V_{ct}^{(4)\rightarrow}(\vec{p'},\vec{p}) &= D_1 q^4 + D_2 k^4 + D_3 q^2 k^2 + D_4 \left(\vec{q}\times\vec{k}\right)^2 \\ &+ \left(D_5 q^4 + D_6 k^4 + D_7 q^2 k^2 + D_8 (\vec{q}\times\vec{k})^2\right)\vec{\sigma}_1.\vec{\sigma}_2 \\ &+ (D_9 q^2 + D_{10} k^2)(-i\vec{S}.(\vec{q}\times\vec{k})) \\ &+ (D_{11} q^2 + D_{12} k^2)(\vec{\sigma}_1.\vec{q})(\vec{\sigma}_2.\vec{q}) \\ &+ (D_{13} q^2 + D_{14} k^2)\left(\vec{\sigma}_1.\vec{k}\right)\left(\vec{\sigma}_2.\vec{k}\right) \\ &+ D_{15}\left(\vec{\sigma}_1.\left(\vec{q}\times\vec{k}\right)\vec{\sigma}_2.\left(\vec{q}\times\vec{k}\right)\right). \end{aligned}$$

The 2PE potential at N^{3}LO, $V_{2\pi}^{(4)}$, is very involved, which is why will not give its expressions here. The N^{3}LO 3PE contributions, $V_{3\pi}^{(4)}$, are negligible.

The two-nucleon system is characterized by large scattering lengths and a shallow bound states (the deuteron), which cannot be calculated by perturbation theory. Therefore, the *NN* potential must be inserted into a Schrödinger or Lippmann-Schwinger (LS) equation to obtain the *NN* amplitude,

$$\hat{T}(\vec{p}',\vec{p}) = \hat{V}(\vec{p}',\vec{p}) + \int d^3p''\hat{V}(\vec{p}',\vec{p}'')\frac{M_N}{p^2 - p''^2 + i\in}\hat{T}(\vec{p}'',\vec{p}),$$

where the definitions

$$\hat{V}(\vec{p}',\vec{p}) \equiv \frac{1}{(2\pi)^3}\sqrt{\frac{M_N}{E_{p'}}}V(\vec{p}',\vec{p})\sqrt{\frac{M_N}{E_{p'}}}$$

and

$$\hat{T}(\vec{p}',\vec{p}) \equiv \frac{1}{(2\pi)^3}\sqrt{\frac{M_N}{E_{p'}}}T(\vec{p}',\vec{p})\sqrt{\frac{M_N}{E_{p'}}},$$

are used, with $E_p \equiv \sqrt{M_N^2 + p^2}$ Iteration of $\hat{V}$ in the LS equation.

$\hat{T}(\vec{p}',\vec{p})=\hat{V}(\vec{p}',\vec{p})+\int d^3p''\hat{V}(\vec{p}',\vec{p}'')\frac{M_N}{p^2-p''^2+i\in}\hat{T}(\vec{p}'',\vec{p})$, requires cutting $\hat{V}$ off for high momenta to avoid infinities. This is consistent with the fact that ChPT is a low-momentum expansion which is valid only for momenta $Q<\Lambda\chi\approx 1$ GeV. Therefore, the potential $\hat{V}$ is multiplied with a regulator function $f(p',p)$,

$$\hat{V}(\vec{p}',\vec{p})\mapsto\hat{V}(\vec{p}',\vec{p})f(p',p)$$

with, for example,

$$f(p',p)=\exp\left[-p'/\Lambda)^{2n}-(p/\Lambda)^{2n}\right]$$

Typical choices for the cutoff parameter Λ that appears in the regulator are $\Lambda\approx 0.5$ GeV$<\Lambda\chi\approx 1$ GeV.

It is pretty obvious that results for the $\hat{T}$ -matrix may then depend sensitively on the regulator and its cutoff parameter, which is undesirable. The removal of such regulator dependence is known as renormalization. Note that renormalizability is crucial for the validity of an EFT. The quantitative chiral *NN* potentials currently in use apply two regularization schemes: In the derivation of the potential, dimensional regularization is used (where the cutoff is taken to infinity), while in the LS equation, Equation $\hat{T}(\vec{p}',\vec{p})=\hat{V}(\vec{p}',\vec{p})+\int d^3p''\hat{V}(\vec{p}',\vec{p}'')\frac{M_N}{p^2-p''^2+i\in}\hat{T}(\vec{p}'',\vec{p})$, the regulator Equation $f(p',p)=\exp\left[-p'/\Lambda)^{2n}-(p/\Lambda)^{2n}\right]$ is applied with a finite cutoff. This scheme has produced useful *NN* potentials, but the way regularization and renormalization is handled is controversial. In spite of almost two decades of research by a large variety of theoretical physicists, there is still no consensus in the community on how to conduct the renormalization of chiral EFT based nuclear forces in a satisfactory way. In this context, the pionless EFT has turned out to be enlightening, since it allows for a more transparent renormalization procedure (because it consists of contacts only and does not include pion loops).

A related unresolved issue is the proper counting of the powers of the low-energy expansion. Notice that the power given in Equation $\nu=-2+2A-2C+2L+\sum\Delta_i$, is the one of the perturbatively calculated potential. However, the quantity of physical relevance is the $\hat{T}$ -matrix, which is obtained by iterating the potential in the LS equation, Equation $\hat{T}(\vec{p}',\vec{p})=\hat{V}(\vec{p}',\vec{p})+\int d^3p''\hat{V}(\vec{p}',\vec{p}'')\frac{M_N}{p^2-p''^2+i\in}\hat{T}(\vec{p}'',\vec{p})$, which may change the power. Also here the pionless theory has been helpful since, due to its simplicity, it allows for analytic solutions of the LS equation revealing the power explicitly.

Three-nucleon Forces

In microscopic calculations of nuclear structure and reactions, the 2NF makes, of course, the largest contribution. However, from *ab-initio* studies it is well-known that certain few-nucleon reactions and nuclear structure issues require 3NFs for their precise microscopic explanation. In short, we need 3NFs. As noted before, an important advantage of the EFT approach to nuclear forces is that it creates two- and many-nucleon forces on an equal footing above in figure.

For a 3NF, we have A=3 and C=1 and, thus, Equation $\nu = -2 + 2A - 2C + 2L + \sum_i \Delta_i$, implies

$$\upsilon = 2 + 2L + \sum_i \Delta_i.$$

We will use this equation to analyze 3NF contributions order by order.

The lowest possible power is obviously ν=2 (NLO), which is obtained for no loops (L=0) and only leading vertices ($\sum_i \Delta_i = 0$). As it turns out, this contribution vanishes.

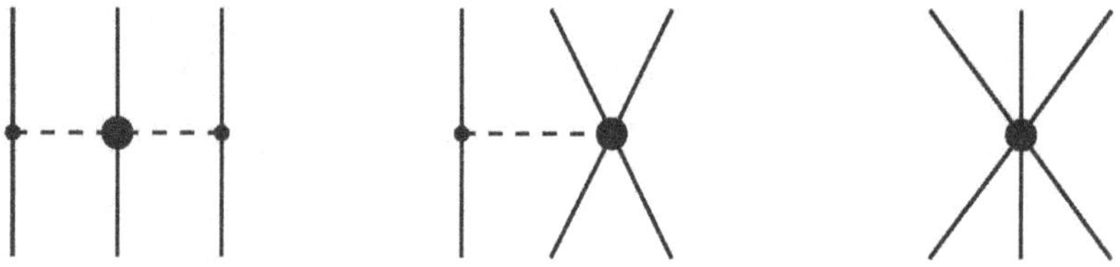

Figure: 3NF at NNLO (ν=3). From left to right: 2PE, 1PE-contact, and contact diagrams.

The first non-vanishing 3NF appears at NNLO (ν=3). The power ν=3 is obtained when there are no loops (L=0) and $\sum_i \Delta_i = 1$, i.e., Δi=1 for one vertex while Δi=0 for all other vertices. There are three topologies which fulfill this condition, known as the 2PE, 1PE, and contact graphs.

The 2PE 3N-potential is derived to be

$$V^{3\mathrm{NF}}_{2\mathrm{PE}} = \left(\frac{g_A}{2f_x}\right)^2 \frac{1}{2} \sum_{i \neq j \neq k} \frac{(\vec{\sigma}_i.\vec{q}_i)(\vec{\sigma}_j.\vec{q}_j)}{(q_i^2 + m_\pi^2)(q_j^2 + m_\pi^2)} F^{ab}_{ijk} \tau_i^a \tau_j^b$$

with $\vec{q}_i \equiv \vec{p}_i - \vec{p}_i$, where $\vec{p}_i$ and $\vec{p}\,'_i$ are the initial and final momenta of nucleon i, respectively, and

$$F^{ab}_{ijk} = \aleph^{ab}\left[-\frac{4c_1 m_\pi^2}{\ddot{u}_{\aleph}^{\ddot{u}}} + \frac{2c_3}{}\vec{q}_i \ddot{u}\vec{q}_j\right] + \frac{c_4}{}\sum_c \in^{abc} {}^{c}_{k} \, {}^{\rightarrow}_{k} \left[\vec{q}_i \times \vec{q}_j\right]$$

There are great similarities between this force and derivations of 2PE 3NFs from conventional meson theory.

The other two 3NF contributions are easily derived by taking the last two terms of the Δ=1 Lagrangian, equation

$$
\begin{aligned}
\hat{\mathcal{L}}^{\Delta=1} = \bar{N}&\left\{\frac{\vec{\nabla}^2}{2M_N} - \frac{ig_A}{4M_N f_\pi}\tau.\left[\vec{\sigma}.\left(\vec{\nabla}\partial_0\pi - \partial_0\pi\,\vec{\nabla}\right)\right]\right\} \\
&-\frac{i}{8M_N f_\pi^2}\tau.\left[\vec{\nabla}.\left(\pi\times\vec{\nabla}\pi\right)-\left(\pi\times\vec{\nabla}\pi\right).\vec{\nabla}\right]\Bigg\}N \\
&+\bar{N}\left[4_{c1m_\pi^2} - \frac{2_{c1}}{f_\pi^2}m_\pi^2\pi^2 + \left(c_2 - \frac{g_A^2}{8M_N}\right)\frac{1}{f_\pi^2}\left(\partial_0\pi.\partial_0\pi\right)\right. \\
&\left.+\frac{c_3}{f_\pi^2}\left(\partial_u\pi.\partial^u\pi\right) - \left(c_4 + \frac{1}{8M_N}\right)\frac{1}{f_\pi^2}\in^{ijk}\in^{abc}\sigma^i\tau^a\left(\partial^j\pi^b\right)\left(\partial^k\pi^c\right)\right]N \\
&-\frac{D}{4f_\pi}\left(\bar{N}N\right)\bar{N}\left[\tau.\left(\vec{\sigma}.\vec{\nabla}\right)\pi\right]N - \frac{1}{2}E\left(\bar{N}N\right)\left(\bar{N}\tau N\right).\left(\bar{N}\tau N\right)+..,
\end{aligned}
$$

into account. The 1PE contribution is

$$
V_{1\text{PE}}^{3\text{NF}} = -D^2\frac{g_A}{8f_x^2}\sum_{i\neq j\neq k}\frac{\vec{\sigma}_j.\vec{q}_j}{q_i^2+m_\pi^2}\left(\tau_i.\tau_j\right)\left(\vec{\sigma}_i.\vec{q}_j\right)
$$

and the 3N contact potential reads

$$
V_{ct}^{3\text{NF}} = E\frac{1}{2}\sum_{j\neq k}\tau_j.\tau_k.
$$

These 3NF terms involve the two new parameters D and E, which do not appear in the 2N problem. There are many ways to pin these two parameters down. Using the triton binding energy and the *nd* doublet scattering length ${}^2a_{nd}$ is one possibility. One may also choose the binding energies of ${}^3H$ and ${}^4He$ or an optimal over-all fit of the properties of light nuclei. Once D and E are fixed, the results for other ${}_3$N, ${}_4$N, etc. observables are predictions.

The 3NF at NNLO has been applied in calculations of few-nucleon reactions, structure of light- and medium-mass nuclei, and nuclear and neutron matter with a good deal of success. Yet, the problem with the underprediction of the analyzing power of nucleon-deuteron and p-^{3}He scattering, which has become known as the 'A_y puzzle', is not resolved by this 3NF. Furthermore, the spectra of light nuclei leave room for improvement. Therefore, 3NFs of higher orders are needed for at least two reasons: to hopefully resolve outstanding problems in microscopic structure and reactions and for consistency with the 2NF (recall that a precise 2NF is of order N^{3}LO).

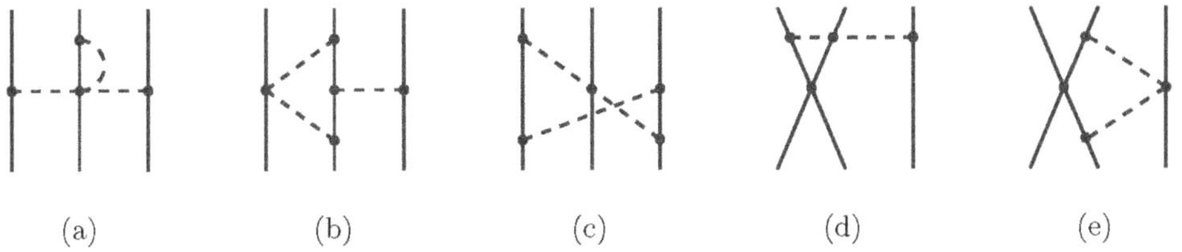

Figure: 3NF at N3LO (ν=4). We show one representative diagram for each of five topologies, which are: (a) 2PE, (b) 2PE-1PE, (c) ring, (d) 1PE-contact, and (e) 2PE-contact.

The next order is N^3|LO (**v**=4), where we are faced with a very large number of loop diagrams in above figure. For those loops, *L* is one and, therefore, all Δ_i have to be zero to ensure v=4. Thus, these one-loop 3NF diagrams can include only leading order vertices, the parameters of which are fixed from *πN* and *NN* analysis. There are five loop topologies. In figure we show one sample diagram for each topology. Note, however, that each topology consists of many diagrams such that the total number of diagrams is between 50 and 100, depending on how the diagrams are represented. Preliminary applications of the 3N potentials derived from these diagrams indicate that the N3LO 3NF is fairly weak and does not solve the *Ay* puzzle.

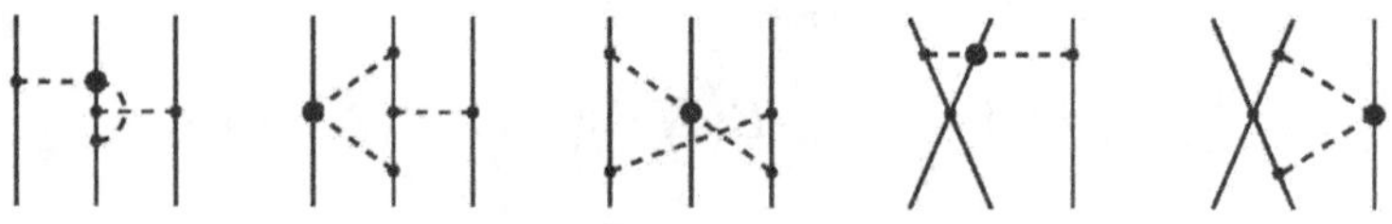

(a) (b) (c) (d) (e)

Figure: 3NF loop contributions at N^4LO (v=5). We show one representative diagram for each of five topologies, which are: (a) 2PE, (b) 2PE-1PE, (c) ring, (d) 1PE-contact, and (e) 2PE-contact.

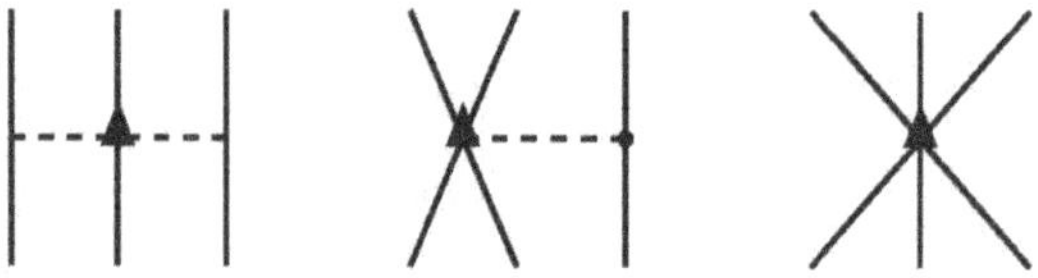

(a) (b) (c)

Figure: 3NF 'tree' graphs at N4LO (v=5) denoted by: (a) 2PE, (b) 1PE-contact, and (c) contact. Solid triangles represent vertices of index Δ_i=3.

Since we are dealing with a perturbation theory, it is natural to turn to the next order when looking for further improvements. Thus, we proceed to order N^4LO (**v**=5). The loop contributions that occur at this order in above figure are obtained by replacing in the N^3LO loops one vertex by a $\Delta_{i=1}$ vertex (with LEC *ci*)}, which is why these loops may be more sizable than the N^4LO loops. Again, there are five loop topologies, each of which consists of many diagrams. In addition, we have three 'tree' topologies above in figure which include a new set of 3N contact interactions [graph (c)]. Contact terms are typically simple (as compared to loop diagrams) and their coefficients are unconstrained (except for naturalness). The N^4LO $_3$NF terms include all possible spin-isospin-momentum structures that a 3NF can have. Thus, there is hope that the 3NF at N4LO may provide the missing pieces in the 3NF puzzle. However, a problem is how to deal with the explosion of 3NF contributions that emerge at N^3LO and N^4LO.

Four-nucleon Forces

For connected (*C*=1) *A*=4 diagrams, Eq. $v=-2+2\ddot{u}-2 \quad +2 \quad +\sum_i \Delta_i$, yields

$$v=4+2L+\sum_i \Delta_i.$$

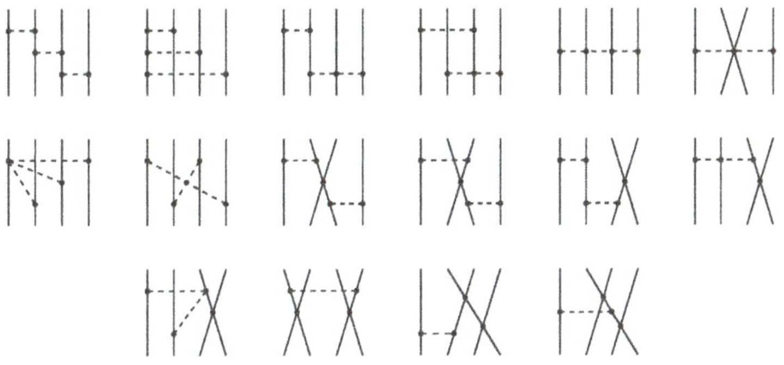

Figure: Leading 4NF at N^3LO.

Therefore, a connected 4NF appears for the first time at v=4 (N^3LO), with no loops and only leading vertices, above in figure. This 4NF includes no new parameters and does not vanish. Some graphs in above in figure appear to be reducible (iterative). Note, however, that these are Feynman diagrams, which are best analyzed in terms of time-ordered perturbation theory. The various time-orderings include also some irreducible topologies (which are, by definition, 4NFs). Or, in other words, the Feynman diagram minus the reducible part of it yields the (irreducible) contribution to the 4NF.

Assuming a good rate of convergence, a contribution of order $(Q/\Lambda_\chi)^4$ is expected to be rather small. Thus, ChPT predicts 4NF to be essentially insignificant, consistent with experience. Still, nothing is fully proven in physics unless we have performed explicit calculations. The leading 4NF has been applied in a calculation of the ^4He binding energy, where it was found to contribute about 0.1 MeV. This is small as compared to the full ^4He binding energy of 28.3 MeV.

Introducing Δ-isobar Degrees of Freedom

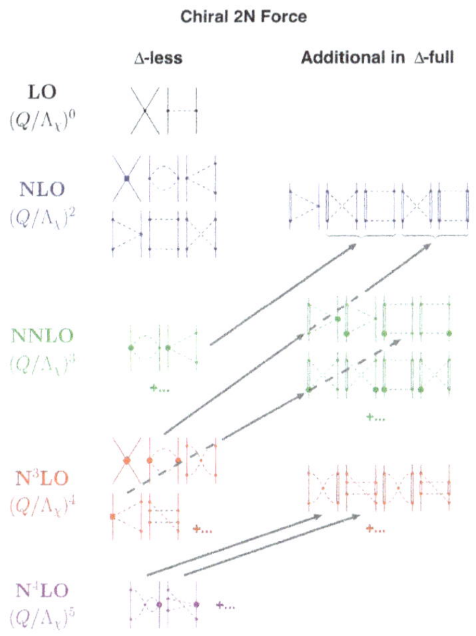

Figure: Chiral 2NF without and with Δ-isobar degrees of freedom. Arrows indicate the shift of strength when explicit Δ's are added to the theory. Note that the Δ-full theory consists of the diagrams involving Δ's *plus* the Δ-less ones.

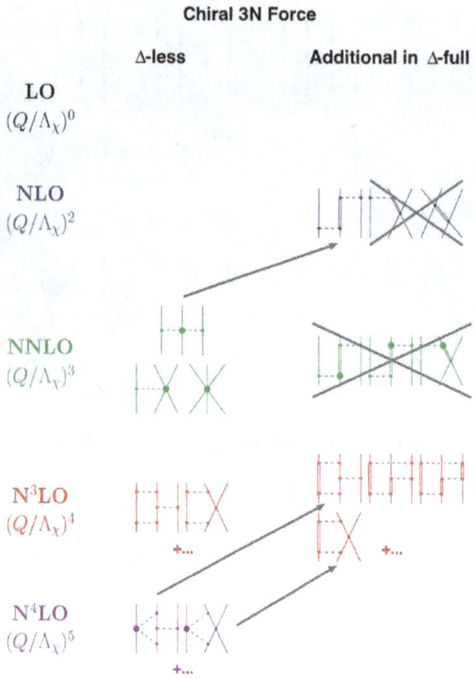

The lowest excited state of the nucleon is the Δ(1232) resonance or isobar (a π-N P-wave resonance with both spin and isospin 3/2) with an excitation energy of $\Delta M = M_\Delta - M_N = 293\,MeV$. Because of its strong coupling to the π-Nsystem and low excitation energy, it is an important ingredient for models of pion-nucleon scattering in the Δ-region and pion production from the two-nucleon system at intermediate energies, where the particle production proceeds prevailingly through the formation of Δ isobars. At low energies, the more sophisticated conventional models for the 2π-exchange contribution to the *NN* interaction include the virtual excitation of Δ's, which in these models accounts for about 50% of the intermediate-range attraction of the nuclear force---as demonstrated by the Bonn potential.

Because of its relatively small excitation energy, it is not clear from the outset if, in an EFT, the Δ should be taken into account explicitly or integrated out as a heavy degree of freedom. If it is included, then $\Delta M \sim m_\pi$ is considered as another small expansion parameter, besides the pion mass and small external momenta. This scheme has become known as the small scale expansion (SSE). Note, however, that this extension is of phenomenological character, since ΔM does not vanish in the chiral limit.

In the chiral EFT discussed so far (also known as the Δ-less theory), the effects due to Δ isobars are taken into account implicitly. Note that the dimension-two LECs, *ci*, have unnaturally large values. The reason for this is that the Δ-isobar (and some meson resonances) contribute considerably to the *ci*---a mechanism that has become known as resonance saturation. Therefore, the explicit inclusion of the Δ [the so-called Δ-full theory), takes strength out of these LECs and moves this strength to a lower order. As a consequence, the convergence of the expansion improves, which is another motivation for introducing explicit Δ-degrees of freedom. In the Δ-less theory, the subleading 2PE and 3PE contributions to the 2NF are larger than the leading ones. The promotion of large contributions by one order in the Δ-full theory fixes this problem.

In analogy to the $NN\pi$ coupling, which is proportional to $g_A/2f_\pi$ and includes one derivative, the $N\Delta\pi$ coupling is proportional to $h_A/2f_\pi$ and includes one derivative. The LECs of the π-N Lagrangian are usually extracted in the analysis of π-N scattering data and clearly come out differently in the Δ-full theory as compared to the Δ-less one. While in the Δ-less theory the magnitude of the LECs c3and c4 is about 3-5 GeV−1, they turn out to be around 1 GeV−1 in the Δ-full theory.

In the 2NF, the virtual excitation of Δ-isobars requires at least one loop and, thus, the contribution occurs first at v=2 (NLO). The consistency between the Δ-full and Δ-less theories has been verified by showing that the contributions due to intermediate Δ-excitations, expanded in powers of $1/\Delta M$, can be absorbed into a redefinition of the LECs of the Δ-less theory. The corresponding shift of the LECs c_3, c_4 is given by

$$c_3 = -2c_4 = -\frac{h_A^2}{9\Delta M}.$$

Using $h_A = 3g_A/\sqrt{2}$ (large Nc value), almost all of c_3 and an appreciable part of c_4 is explained by the Δ resonance.

Several studies have confirmed that a large amount of the intermediate-range attraction of the 2NF is shifted from NNLO to NLO with the explicit introduction of the Δ-isobar. However, it is also found that the NNLO 2PE potential of the Δ-less theory provides a very good approximation to the NNLO potential in the Δ-full theory.

The Δ isobar also changes the 3NF scenario. The leading 2PE 3NF is promoted to NLO. In the Δ-full theory, this term has the same mathematical form as the corresponding term in the Δ-less theory,

equation $V_{2\text{PE}}^{3\text{NF}} = \left(\frac{g_A}{2f_x}\right)^2 \frac{1}{2}\sum_{i\neq j\neq k} \frac{(\vec{\sigma}_i.\vec{q}_i)(\vec{\sigma}_j.\vec{q}_j)}{(q_i^2+m_\pi^2)(q_j^2+m_\pi^2)} F_{ijk}^{ab}\tau_i^a\tau_j^b$ and

$F_{ijk}^{ab} = \delta^{ab}\left[-\frac{4c_1m_\pi^2}{f_\pi^2} + \frac{2c_3}{f_\pi^2}\vec{q}_i.\vec{q}_j\right] + \frac{c_4}{f_\pi^2}\sum_c \epsilon^{abc}\,\tau_k^c\vec{\sigma}_k.\left[\vec{q}_i\times\vec{q}_j\right]$., provided one chooses c_1=0 and c_3, c_4 according to Eq.~($c_3 = -2_{c_4} = -\frac{h_A^2}{9\Delta M}$.).

Note that the other two NLO 3NF terms involving Δ's vanish as a consequence of the antisymmetrisation of the 3N states. The Δ contributions to the 3NF at NNLO vanish at this order, because the subleading $N\Delta\pi$ vertex contains a time-derivative, which demotes the contributions by one order. However, substantial 3NF contributions are expected at N^3LO from one-loop diagrams with one, two, or three intermediate Δ-excitations, which correspond to diagrams of order N^4LO, N^5LO, and N^6LO, respectively, in the Δ-less theory.

To summarize, the inclusion of explicit Δ degrees of freedom does certainly improve the convergence of the chiral expansion by shifting sizable contributions from NNLO to NLO. On the other hand, at NNLO the results for the Δ-full and Δ-less theory are essentially the same. Note that the Δ-full theory consists of the diagrams involving Δ's plus all diagrams of the Δ-less theory. Thus, the Δ-full theory is much more involved. Moreover, in the Δ-full theory, $1/MN$ 2NF corrections appear at NNLO, which were found to be uncomfortably large. Thus, it appears that up to NNLO, the Δ-less theory is more manageable.

The situation could, however, change at N^3LO where potentially large contributions enter the picture. It may be more efficient to calculate these terms in the Δ-full theory, because in the Δ-less theory they are spread out over N^3LO, N^4LO and, in part, $ü^5$. These higher order contributions are a crucial test for the convergence of the chiral expansion of nuclear forces and represent a challenging topic for future research.

Baryon-baryon Interactions

All baryons interact strongly with each other. Therefore, besides interactions between nucleons, there are many more strong baryon-baryon interactions. Traditionally, one focus has been the forces between nucleons and hyperons (strange baryons) and hyperons and hyperons. Furthermore, the interaction between a baryon and an anti-baryon has drawn considerable interest, for which the nucleon-antinucleon interaction is the most studied example. As in the case of the nucleon-nucleon interaction, the approaches that have been tried to explain the baryon-(anti)baryon interactions include: phenomenology, meson theory, quark models, lattice QCD, and effective field theory. It is not the purpose of this article to discuss those other baryon-baryon interactions, but it is worthwhile pointing out that due to the quark sub-structure of hadrons, all baryon-baryon interactions are related. Thus, the nucleon-nucleon interaction discussed in this part is not an isolated object and should be viewed in the context of all baryonic interactions. A description of all baryon-baryon interactions that is consistent with all relevant underlying symmetries is a challenging subject of contemporary research. For more information on this topic, we like to refer the interested reader to the literature. Hyperon-nucleon interactions are reviewed by in the meson picture and by in chiral EFT. For nucleon-antinucleon for a chiral EFT approach.

Elastic and Inelastic Scattering

Elastic scattering occurs when two or more particles collide without any loss of energy. This means that while the directions of the particles may change, the total kinetic energy of the system, or movement energy, is always conserved. The term elastic scattering is typically used in particle physics, which is the study of microscopic particles, but an elastic collision can also take place between macroscopic objects. An inelastic collision occurs when energy is lost during the collision.

The term elastic scattering comes from scattering theory, which is a set of rules and equations which describe how particles and waves interact. In the macroscopic world, when two objects collide it is usually through a physical collision. In particle physics, however, the objects may collide through other forces, including electromagnetic collisions. An elastic collision can occur between any objects and in any type of collision.

Elastic scattering is very important in particle physics. When electrons collide with other particles, for example, the collision is elastic as no energy is lost. This is known as Rutherford scattering and is a phenomenon that led to the discovery of the structure of the atom.

In the macroscopic or physical world, it is exceptionally unlikely for a true elastic collision to occur between two large objects. This is due to surrounding forces as well as the vibrations that occur inside large objects. There are some situations, however, where a collision can be approximated as elastic. This is useful as it allows for the predicted speed and direction of two objects after a collision to be estimated using simpler methods.

A common example of elastic scattering in the physical world is the collision of two billiard balls. Although a small amount of energy will be lost in this collision due to friction, this is small enough to be negligible. When two billiard balls collide, the second ball gains almost exactly the amount of energy that the first ball loses, so the total kinetic energy of the system is conserved.

Inelastic Scattering

In chemistry, nuclear physics, and particle physics, inelastic scattering is a fundamental scattering process in which the kinetic energy of an incident particle is not conserved (in contrast to elastic scattering). In an inelastic scattering process, some of the energy of the incident particle is lost or increased. Although the term is historically related to the concept of inelastic collision in dynamics, the two concepts are quite distinct; inelastic collision in dynamics refers to processes in which the total macroscopic kinetic energy is not conserved. In general, scattering due to inelastic collisions will be inelastic, but, since elastic collisions often transfer kinetic energy between particles, scattering due to elastic collisions can also be *in*elastic, as in Compton scattering.

Electrons

When an electron is the incident particle, the probability of inelastic scattering, depending on the energy of the incident electron, is usually smaller than that of elastic scattering. Thus in the case of gas electron diffraction, reflection high-energy electron diffraction (RHEED), and transmission electron diffraction, because the energy of the incident electron is high, the contribution of inelastic electron scattering can be ignored. Deep inelastic scattering of electrons from protons provided the first direct evidence for the existence of quarks.

Photons

When a photon is the incident particle, there is an inelastic scattering process called Raman scattering. In this scattering process, the incident photon interacts with matter (gas, liquid, and solid) and the frequency of the photon is shifted to red or blue. A red shift can be observed when part of the energy of the photon is transferred to the interacting matter, where it adds to its internal energy in a process called Stokes Raman scattering. The blue shift can be observed when internal energy of the matter is transferred to the photon; this process is called anti-Stokes Raman scattering.

Inelastic scattering is seen in the interaction between an electron and a photon. When a high-energy photon collides with a free electron and transfers energy, the process is called Compton scattering. Furthermore, when an electron with relativistic energy collides with an infrared or visible photon, the electron gives energy to the photon. This process is called inverse Compton scattering.

Neutrons

Neutrons undergo many types of scattering, including both elastic and inelastic scattering. Whether elastic or inelastic scatter occurs is dependent on the speed of the neutron, whether fast or thermal, or somewhere in between. It is also dependent on the nucleus it strikes and its neutron cross section. In inelastic scattering, the neutron interacts with the nucleus and the kinetic energy of the system is changed. This often activates the nucleus, putting it into an excited, unstable, short-lived energy state which causes it to quickly emit some kind of radiation to bring it back down to a stable or ground state. Alpha, beta, gamma, and protons may be emitted. Particles scattered in this type of nuclear reaction may cause the nucleus to recoil in the other direction.

Molecular Collisions

Inelastic scattering is common in molecular collisions. Any collision which leads to a chemical reaction will be inelastic, but the term inelastic scattering is reserved for those collisions which do *not* result in reactions. There is a transfer of energy between the translational mode (kinetic energy) and rotational and vibrational modes.

If the transferred energy is small compared to the incident energy of the scattered particle, one speaks of quasielastic scattering.

Strong Nuclear Force

Strong nuclear force is about 100 times stronger than electromagnetism. These forces is also known as strong interactions.

Strong nuclear forces can be applied in two aspects: One is on Larger Scale and other on Lower Scale.

- On larger scale of about 1 to 3 fm distance, it is the force that binds the nucleons together to form nuclei.
- On smaller scale of about less than 0.8 fm distance, it is the force that binds quarks together to form nucleons that is protons and neutrons and also other particles like hadrons.

Strong Nuclear Force Examples

Strong nuclear forces help in holding sub atomic particles of protons together and also the nucleons together at larger scale.

- Strong nuclear force leads to release of energy when heat is generated in Nuclear Power Plant to generate steam for generating electricity.
- Energy is released when a Nuclear Weapon detonates which is due to strong nuclear forces.

A natural idea now was to search for a mechanism like the one in electromagnetism to mediate the strong force. Already in 1935 Hideki Yukawa proposed a field theory for the strong interaction where the mediating field particle was to be called a meson.

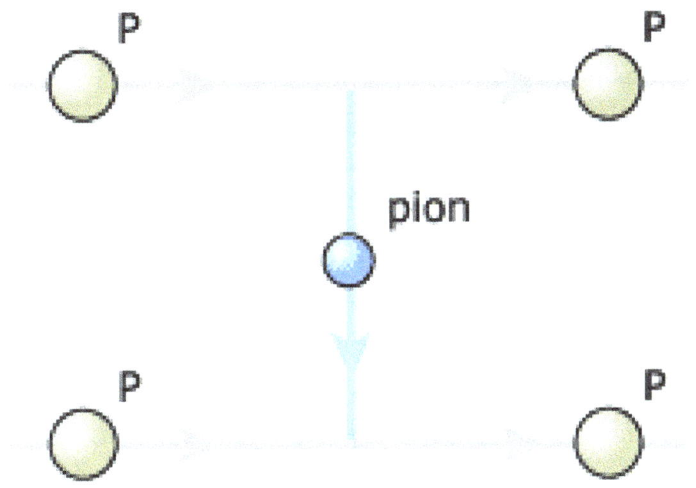

However, there is a significant difference between the strong force and the electromagnetic one in that the strong force has a very short range (typically the nuclear radius). This is the reason why it has no classical counterpart and hence had not been discovered in classical physics. Yukawa solved this problem by letting the meson have a mass.

Theory for Strong Interactions

A remarkable feature of the SLAC experiments that verified the existence of quarks was 'scaling'. The cross sections for the deep inelastic scattering of electrons on protons depended on fewer kinematical variables for higher energies. The cross sections scaled. Richard Feynman explained it by assuming that the protons consisted of point-like constituents. To explain scaling these constituents must have a coupling strength that decreases with energy, opposite to the case of QED. This was called 'asymptotic freedom'. It was quite difficult to believe that a quantum field theory could be asymptotically free since the energy dependence of the coupling constant is due to the screening from pairs of virtual particles. Relativistic quantum mechanics allow for such pairs if they do not live too long. This is due to Heisenberg's uncertainty principle and the fact that energy is the same as mass according to Einstein's famous formula.

Asymptotic freedom must mean that the quark charges are antiscreened, which as said was hard to believe to exist in a quantum field theory. for a non-abelian gauge field theory the requirement of asymptotic freedom is satisfied if there are not too many quarks. The key to the solution was that the vector particles mediating the force, the gluons, do indeed antiscreen. This can be understood since the charges of the quarks and the gluons, the "colour charges" satisfy more complicated relations than the simpler electric charges. There are three different colours and their anticolours. While the quarks have a colour charge, the gluons have a colour and an anticolour charge. Hence virtual gluons can line up with charges screening each other while the strength of the field increases.

The discovery of asymptotic freedom opened up for a non-abelian gauge field theory for the interactions among quarks and it was called Quantum Chromodynamics, QCD. Over the years this theory has been very successfully tested at the large accelerators and it is now solidly established as the theory of the strong interactions.

Weak Nuclear Force

The weak force is one of the four fundamental forces that govern all matter in the universe (the other three are gravity, electromagnetism and the strong force). While the other forces hold things together, the weak force plays a greater role in things falling apart, or decaying.

The weak force, or weak interaction, is stronger than gravity, but it is only effective at very short distances. It acts on the subatomic level and plays a crucial role in powering stars and creating elements. It is also responsible for much of the natural radiation present in the universe.

Italian physicist Enrico Fermi devised a theory in 1933 to explain beta decay, which is the process by which a neutron in a nucleus changes into a proton and expels an electron, often called a beta particle in this context. "He defined a new type of force, the so-called weak interaction, that was responsible for decay, and whose fundamental process was transforming a neutron into a proton, an electron and a neutrino," which was later determined to be an anti-neutrino.

Fermi originally thought that this involved what amounted to a zero-distance or adhesive force whereby the two particles actually had to be touching for the force to work. It has since been shown that the weak force is actually an attractive force that works at an extremely short range of about 0.1 percent of the diameter of a proton.

Standard Model

The weak force is part of the reigning theory of particle physics, the Standard Model, which describes the fundamental structure of matter using an "elegant series of equations," Under the Standard Model, elementary particles — that is, those that cannot be split up into smaller parts — are the building blocks of the universe.

One of these particles is the quark. Scientists haven't seen any indication that there is anything smaller than a quark, but they're still looking. There are six types, or "flavors," of quarks: up, down, strange, charm, bottom and top (in ascending order by mass). In different combinations, they form many varied species of the subatomic particle zoo. For example, protons and neutrons, the "big" particles of an atom's nucleus, each consist of bundles of three quarks. Two ups and a down make a proton; an up and two downs make a neutron. Changing the flavor of a quark can change a proton into a neutron, thus changing the element into a different one.

Another type of elementary particle is the boson. These are force-carrier particles that are made up of bundles of energy. Photons are one type of boson; gluons are another. Each of the four forces results from the exchange of force-carrier particles. The strong force is carried by the gluon, while the electromagnetic force is carried by the photon. The graviton is theoretically the force-carrying particle of gravity, but it has not been found yet.

W and Z bosons

The weak force is carried by the W and Z bosons. These particles were predicted by Nobel laureates Steven Weinberg, Sheldon Salam and Abdus Glashow in the 1960s, and discovered in 1983 at CERN.

W bosons are electrically charged and are designated by their symbols: W^+ (positively charged) and W^- (negatively charged). The W boson changes the makeup of particles. By emitting an electrically charged W boson, the weak force changes the flavor of a quark, which causes a proton to change into a neutron, or vice versa. This is what triggers nuclear fusion and causes stars to burn,. The burning creates heavier elements, which are eventually thrown into space in supernova explosions to become the building blocks for planets, along with plants, people and everything else on Earth.

The Z boson is neutrally charged and carries a weak neutral current. Its interaction with particles is hard to detect. Experiments to find W and Z bosons led to a theory combining the electromagnetic force and the weak force into a unified "electroweak" force in the 1960s. However, the theory required the force-carrying particles to be massless, and scientists knew that the theoretical W boson had to be heavy to account for its short range. theorists accounted for the W's mass by introducing an unseen mechanism dubbed the Higgs mechanism, which calls for the existence of a Higgs boson. the world's largest atom smasher observed a new particle "consistent with the appearance of a Higgs boson."

Beta Decay

The process in which a neutron changes into a proton and vice versa is called beta decay. "Beta decay occurs when, in a nucleus with too many protons or too many neutrons, one of the protons or neutrons is transformed into the other."

Beta decay can go in one of two ways, according to the LBL. In beta *minus* decay, sometimes annotated as β^- decay, a neutron decays into a proton, an electron and an antineutrino. In beta *plus*decay, sometimes annotated as β^+ decay, a proton decays into a neutron, a positron and a neutrino. One element can change into another element when one of its neutrons spontaneously changes into a proton through beta minus decay or when one of its protons spontaneously changes into a neutron through beta plus decay.

Electron Capture

Protons can also turn into neutrons through a process called electron capture, or K-capture. When there is an excess number of protons relative to the number of neutrons in a nucleus, an electron, usually from the innermost electron shell, will seem to fall into the nucleus. "In electron capture, an orbital electron is captured by the parent nucleus, and the products are the daughter nucleus and a neutrino." The atomic number of the resulting daughter nucleus is reduced by 1, but the total number of protons and neutrons remains the same.

Nuclear Fusion

The weak force plays an important role in nuclear fusion, the reaction that powers the sun and thermonuclear (hydrogen) bombs. The first step in hydrogen fusion is to smash two protons together with enough energy to overcome the mutual repulsion they experience due to the electromagnetic force. If the two particles can be brought close enough to each other, the strong force can bind them together. This creates an unstable form of helium (^2He), which has a nucleus with two protons, as opposed to the stable form of helium ^4He, which has two protons and two neutrons.

The next step is where the weak force comes into play. Because of the overabundance of protons, one of the pair undergoes beta decay. After that, other subsequent reactions, including the intermediate formation and fusion of ^3He, eventually form stable ^4He.

Weak Nuclear Force is of two Types

- Charged current Nuclear Force and
- Neutral current Nuclear Force.

Charged current nuclear force is so called because this force is carried by electric charge carriers i.e. W+ and W- boson particles.

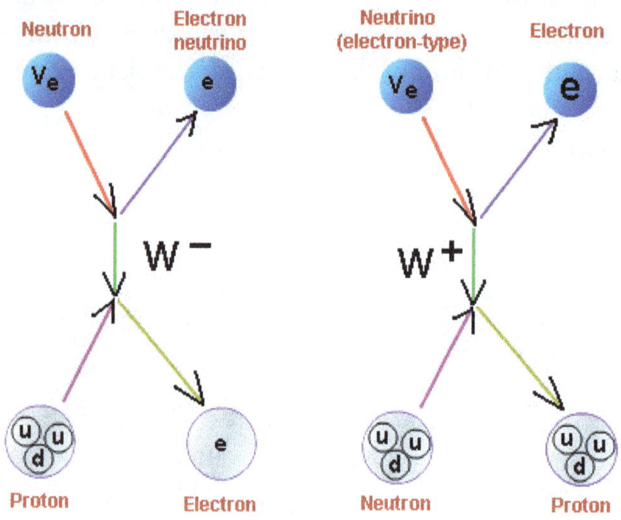

Neutral current nuclear force is carried by neutral particles i.e. Z boson particles.

These Forces Occurs in many Reactions Namely

- Radioactive decay
- Beta decay
- Burning of sun
- Initiating the process of hydrogen fusion in stars.
- In production of deuterium
- Radiocarbon dating and
- Radio luminescence.

Weak Nuclear Force

Weak Nuclear Force is involved in many phenomenon in nature:

Charged Current Nuclear Force:

- A charged electron or muon (lepton) with -1 charge absorbs W+ particle of charge +1 and is converted into neutrino of charge zero. Type of neutrino will be the same as that of lepton.
- A down quark can be converted into up type quark by releasing W- particle.
- An up type quark is converted into down type by absorbing W- particle or emitting W+ boson particle.
- A W boson particle is not stable. Therefore it will decompose or decay in a short time.

Current Nuclear Particle: A quark or lepton can emit or absorb Z boson particle of zero charge. Z boson is also unstable hence decomposes very soon.

It can transform a neutron into proton and a proton into neutron.

Neutron (n) $\rightarrow$ proton (p) + electron (e^-) + anti neutrino ($\bar{v}$) - - - - -> Electron decay
Proton (p) $\rightarrow$ neutron (n) + positron (e^+) + neutrino (v) - - - - -> Positron decay.

It is responsible for radio active decay of particles. It initiates the process of hydrogen fusion at stars. It is also responsible for beta decay which is a form of radioactivity. It is also responsible for production of deuterium and helium from hydrogen which helps in burning of sun and powers the sun's thermonuclear energy. It is also possible to perform radio carbon dating with weak nuclear force since C – 14 decays into N – 14 due to weak nuclear force. Weak nuclear force can also create radio luminescence. It also helps in the formation of heavy nucleus.

Electron Interaction

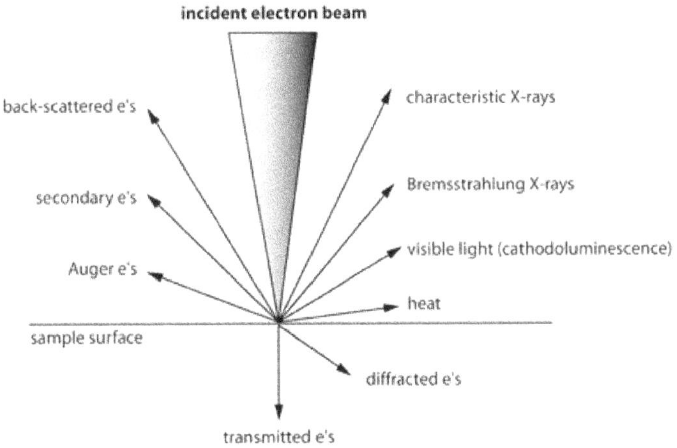

Figure: Types of interactions between electrons and a sample.

Electrons accelerated onto a material result in a number of interactions with the atoms of the target sample. Accelerated electrons can pass through the sample without interaction, undergo elastic scattering and can be inelastically scattered figure. Elastic and inelastic scattering result in a number of signals that are used for imaging, quantitative and semi-quantitative information of the target sample and generation of an X-ray source. Typical signals used for imaging include secondary electrons (SE), backscattered electrons (BSE), cathodoluminescence (CL), auger electrons and characteristic X-rays. Quantitative and semiquantitative analyses of materials as well as element mapping typically utilize characteristic X-rays. Bremsstrahlung (continuum) radiation is a continuous spectrum of X-rays from zero to the energy of the electron beam, and forms a background in which characteristic X-ray must be considered. Further, X-rays generated from a specific target material are used as the roughly fixed-wavelength energy source for X-ray diffraction (XRD) and X-ray fluorescence (XRF) investigations.

Where an electron beam impinges on a sample, electron scattering and photon- and X-ray-production develops in a volume (the electron interaction volume) that is dependent on several factors. These include:

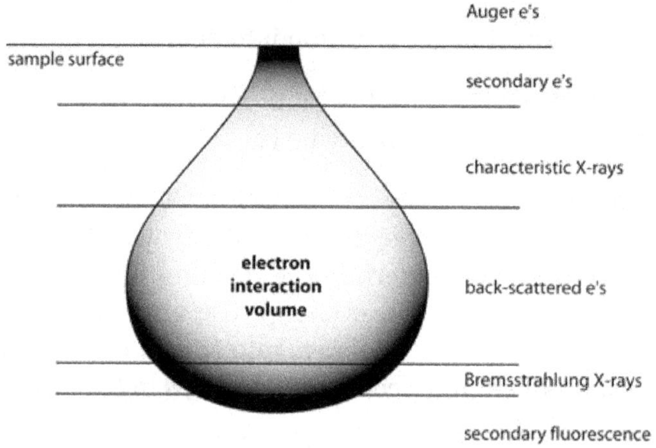

Figure: Electron interaction volume within a sample.

- The energy of the incident beam (accelerating potential) increases the interaction volume, but decreases the elastic scattering (e.g. backscattering).
- The interaction volume decreases as a function of the mean atomic weight.
- Smaller and more asymmetric interaction volumes develop in samples tilted relative to the impinging electron beam.

Each of the signals used for imaging or X-ray generation is generated from different electron interaction volumes and, in turn, each of the signals has different imaging or analytical resolution. Auger and Secondary images have the best imaging resolution, being generated in the smallest volume near the surface of the sample. Backscattered electrons are generated over a larger volume resulting in images of intermediate resolution. Cathodoluminescence is generated over the largest volume, even larger than Bremsstrahlung radiation, resulting in images with the poorest resolution.

Interaction of Proton with Matter

Below figure illustrates several mechanisms by which a proton interacts with an atom or nucleus: Coulombic interactions with atomic electrons, Coulombic interactions with the atomic nucleus, nuclear reactions, and Bremsstrahlung. To a first-order approximation, protons continuously lose kinetic energy via frequent inelastic Coulombic interactions with atomic electrons. Most protons travel in a nearly straight line because their rest mass is 1832 times greater than that of an electron. In contrast, a proton passing close to the atomic nucleus experiences a repulsive elastic Coulombic interaction which, owing to the large mass of the nucleus, deflects the proton from its original straight-line trajectory. Non-elastic nuclear reactions between protons and the atomic nucleus are less frequent but, in terms of the fate of an individual proton, have a much more profound effect. In a nuclear reaction, the projectile proton enters the nucleus; the nucleus may emit a proton, deuteron, triton, or heavier ion or one or more neutrons. Finally, proton Bremsstrahlung is theoretically possible, but at therapeutic proton beam energies this effect is negligible. Table summarizes the proton interaction types, interaction targets, principal ejectiles, influence on the proton beam, and dosimetric manifestations. We review these interaction mechanisms, except proton Bremsstrahlung, in the following sections.

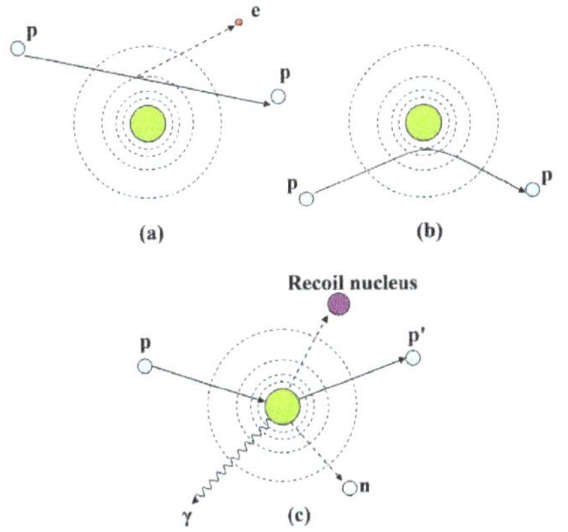

Figure: Schematic illustration of proton interaction mechanisms: (a) energy loss via inelastic Coulombic interactions, (b) deflection of proton trajectory by repulsive Coulomb elastic scattering with nucleus, (c) removal of primary proton and creation of secondary particles via non-elastic nuclear interaction (p: proton, e: electron, n: neutron, γ : gamma rays)

Energy Loss Rate

The energy loss rate of ions, or linear stopping power, is defined as the quotient of dE and dx, where E is the mean energy loss and x is the distance. It is frequently more convenient to express the energy loss rate in a way that is independent of the mass density; the mass stopping power is defined as

Interaction type	Interaction target	Principal ejectiles	Influence on projectile	Dosimetric manifestation
Inelastic Coulomb scattering	Atomic electrons	Primary proton, ionization electrons	Quasi-continuous energy loss	Energy loss determines range in patient
Elastic Coulomb scattering	Atomic nucleus	Primary proton, recoil nucleus	Change in trajectory	Determines lateral penumbral sharpness
Non-elastic nuclear reactions	Atomic nucleus	Secondary protons and heavier ions, neutrons, and gamma rays	Removal of primary proton from beam	Primary fluence, generation of stray neutrons, generation of prompt gammas for *in vivo* interrogation
Bremsstrahlung	Atomic nucleus	Primary proton, Bremsstrahlung photon	Energy loss, change in trajectory	Negligible

Table: Summary of proton interaction types, targets, ejectiles, influence on projectile, and selected dosimetric manifestations.

$$\frac{S}{\rho}=-\frac{dE}{\rho dx},$$

where ρ is the mass density of the absorbing material. Please note that stopping power is defined for a beam, not a particle.

The energy loss rate may be described by several mathematical formulae. The simplest, yet still remarkably accurate, formula is based on the Bragg–Kleeman (BK) rule, which was originally derived for alpha particles, and is given by:

$$\frac{S}{\rho}=-\frac{dE}{\rho dx}\approx-\frac{E^{1-p}}{\rho\alpha p},$$

where ρ is the mass density of the material, α is a material-dependent constant, E is the energy of the proton beam, and the exponent p is a constant that takes into account the dependence of the proton's energy or velocity. Values of α and p may be obtained by fitting to either ranges or stopping power data from measurements or theory.

A more physically complete theory, developed by Bohr, is based on calculation of the momentum impulse of a stationary, unbound electron and the impact parameter. A more accurate formula, attributed to Bethe and Bloch, takes into account quantum mechanical effects and is given by

$$\frac{S}{\rho}=-\frac{dE}{\rho dx}=4\pi N_A r_e^2 m_e c^2\frac{Z}{A}\frac{z^2}{\beta^2}\left[\mathrm{In}\frac{2m_e c^2\gamma^2\beta^2}{I}-\beta^2-\frac{\delta}{2}-\frac{C}{Z}\right],$$

where N_A is Avogadro's number, r_e is the classical electron radius, m_e is the mass of an electron, z is the charge of the projectile, z is the atomic number of the absorbing material, A is the atomic weight of the absorbing material, c is speed of light, $\beta=v/c$ where v is the velocity of the projectile, $\gamma=\left(1-\beta^2\right)^{-ü}$, I is the mean excitation potential of the absorbing material, δ is the density corrections arising from the shielding of remote electrons by close electrons and will result in a reduction of energy loss at higher energies, and C is the shell correction item, which is important only for low energies where the particle velocity is near the velocity of the atomic electrons. The two correction terms in the Bethe–Bloch equation involve relativistic theory and quantum mechanics and need to be considered when very high or very low proton energies are used in calculations. Figure plots proton stopping power as a function of proton energy in water calculated by using

equation $\frac{S}{\rho}=-\frac{dE}{\rho dx}=4\pi N_A r_e^2 m_e c^2\frac{Z}{A}\frac{z^2}{\beta^2}\left[\mathrm{In}\frac{2m_e c^2\gamma^2\beta^2}{I}-\beta^2-\frac{\delta}{2}-\frac{C}{Z}\right]$, at high proton energies (above about 1 MeV) and other methods (not presented) at lower energies.

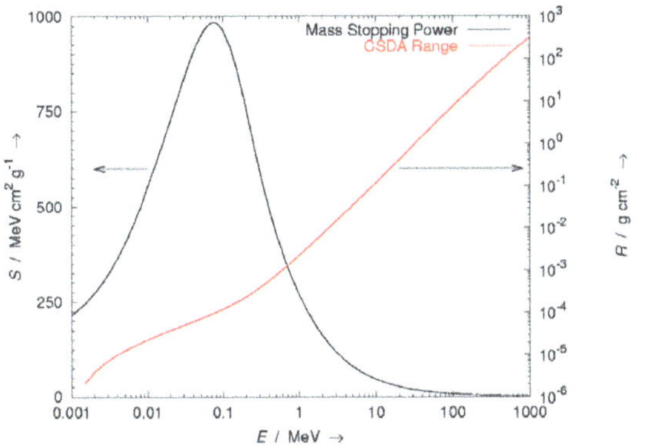

Figure Mass stopping power (S) versus ion energy (E) for protons in liquid water. The corresponding range (R), calculated using the plotted S values and on the assumption of the CSDA, is also plotted.

It is instructive to observe in equation $\frac{S}{\rho}=-\frac{dE}{\rho dx}=4\pi N_A r_e^2 m_e c^2 \frac{Z}{A}\frac{z^2}{\beta^2}\left[\ln\frac{2m_e c^2\gamma^2\beta^2}{I}-\beta^2-\frac{\delta}{2}-\frac{C}{Z}\right]$, how the projectile's characteristics govern its energy loss rate: energy loss is proportional to the inverse square of its velocity ($1/v^2$ classically and $1/\beta^2$ relativistically) and the square of the ion charge (z = 1 for protons), and there is no dependence on projectile mass. Similarly,

equation $\frac{S}{\rho}=-\frac{dE}{\rho dx}=4\pi N_A r_e^2 m_e c^2 \frac{Z}{A}\frac{z^2}{\beta^2}\left[\ln\frac{2m_e c^2\gamma^2\beta^2}{I}-\beta^2-\frac{\delta}{2}-\frac{C}{Z}\right]$, reveals that the absorber material can also strongly influence the energy loss rate. Specifically, the linear stopping power is directly proportional to the mass density. It is equivalent, but perhaps more physically meaningful, to state that the linear stopping power is proportional to the density of electrons in the absorber $\left(N_A \rho Z / A\right)$, because the energy loss occurs by Coulombic interactions between the proton and atomic electrons. Z/A varies by only about 16%, from 0.5 for biologic elements such as carbon and oxygen to 0.42 for high-Z beamline components, such as lead. Hydrogen is an obvious exception to this; fortuitously, the concentration of hydrogen in the human body is low (only about 10%) and nearly uniform throughout the body. The stopping power also depends on a material's I value, and the I value depends in a monotonic way on the Z of the absorber, varying from about 19eV for hydrogen to about 820eV for lead. However, the stopping power goes with the logarithm of I^{-1} value, so the dependence is diminished. Hence, putting these various dependencies in perspective, it is clear that the proton energy loss rate in the human body depends most strongly on the material density, which can vary by about three orders of magnitude, from air in the lung to cortical bone, and the ion velocity, which can cause the linear stopping power in water to vary by about a factor of 60 for proton energies between 1 and 250MeV.

The stopping power theory for protons and heavier ions was reviewed by Ziegler and in Report 49 of the International Commission on Radiation Units and Measurement. Evaluated stopping power and range tables may be conveniently calculated with the SRIM code ('Stopping and Ranges of Ions in Matter,' computer program).

Thus far, we have described proton energy loss in an approximate way on the assumptions that a proton loses energy along a 2D line trajectory and that energy loss is continuous. Absorption of this same energy, however, occurs in a 3D volume. Furthermore, the ionization track of a proton has an irregular 3D structure caused by random fluctuations in the location and size of primary ionization events. This is caused mainly by proton-produced recoil electrons, some of which are sufficiently energetic to create small spur tracks of ionization emanating from the main track. Because the electrons are very much lighter than the protons, each interaction can reduce the proton energy only a little. The maximum possible energy transfer in a collision of an ion of mass m with an unbound stationary electron is

$$\Delta_1^{\max} = \frac{2m_e c^2\beta^2}{1-\beta^2}\left[1+2\frac{m_e}{M}\frac{1}{\sqrt{1-\beta^2}}+\left(\frac{m_e}{M}\right)^2\right],$$

where m_e is the mass of an electron, M is the mass of the target material, c is the speed of light, and $\beta = v/c$ where v is the velocity of the projectile.

Even for very energetic protons, the secondary electrons do not acquire enough energy to travel more than a few millimeters from the proton track. For example, at 200 MeV proton energy, the maximum secondary electron energy is around 500 keV, which corresponds to an electron range of approximately 2 mm in water. The probability of producing secondary electrons may be calculated with various total or differential cross-sections; these were reviewed in ICRU Report 55. Track structure models may be used to estimate the radial properties of ions, although this has not found common application in clinical proton therapy. Regardless of the calculation method used, the spatial characteristics of secondary charged particles should, in principle, be taken into account near material interfaces (e.g. buildup effects in transmission beam monitoring instruments, skin, air–tumor interfaces in the lung) and in cases where the radiation quality is of interest (e.g. microdosimetric and nanodosimetric characterization of individual dose deposition events).

Finally, we note that in proton therapy water is considered an excellent tissue substitute because of its similar density, effective Z/A, and other properties. Furthermore, proton energy loss and residual range in various materials are often expressed in terms of their water-equivalent values.

Range

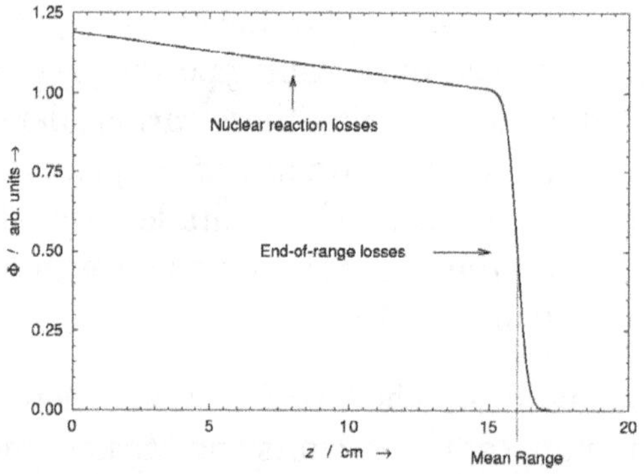

Figure: Relative fraction of the fluence Φ in a broad beam of protons remaining as a function of depth z in water. The gradual depletion of protons from entrance to near the end of range is caused by removal of protons from nuclear reactions. The rapid falloff in the number of protons near the end of range is caused by ions running out of energy and being absorbed by the medium. The sigmoid shape of the distal falloff is caused by range straggling or by stochastic fluctuations in the energy loss of individual protons

Range is defined as the depth at which half of protons in the medium have come to rest, as shown in the range-number curve plotted in below figure. There are small variations in the energy loss of individual protons (an effect called range straggling). Consequently, the range is inherently an average quantity, defined for a beam and not for individual particles. By convention, this means half of the protons incident on the absorber are stopped, although in some cases this is taken instead to mean half of the protons survived to near the end of range (i.e. neglecting protons removed by nuclear reactions).

The path of most protons in matter is a nearly straight line. On average, the proton's pathlength is very nearly equal to its projected pathlength and range. This simple but important fact renders many proton range calculations tractable with relatively simple numerical or analytical approaches.

Let us first consider a simple numerical calculation of proton beam range. We use proton stopping power data and perform a 1D proton pathlength transport calculation on the assumptions that the ions travel only straight ahead (negligible lateral scattering) and that the protons lose energy in a continuous matter. (These assumptions are reasonable for many clinical calculations, but we examine then relax these assumptions in later sections.) In this case, the range (R) may be calculated as:

$$R(E)=\int_0^E\left(\frac{dE'}{dx}\right)^{-1}dE'\approx\sum_0^E\left(\frac{dE'}{dx}\right)^{-1}\Delta E',$$

where E is the ion's initial kinetic energy. The summation denotes that the continuous transport is approximated by calculations of discrete steps. In fact, as discussed above, this equation actually gives the pathlength, which is an excellent approximation of range in most clinical situations. In above figure plots proton range in water calculated by using above equation.

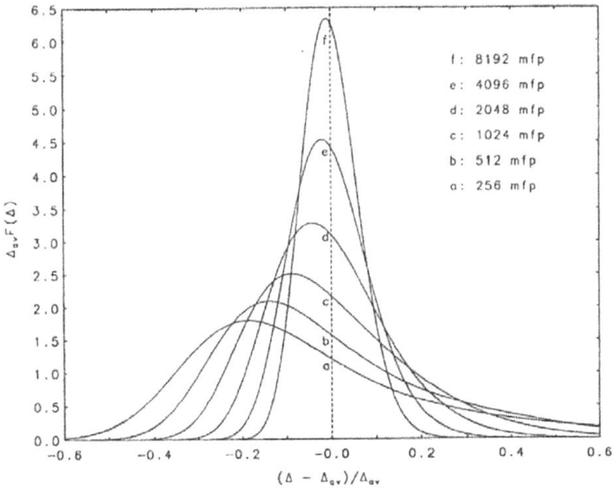

Figure: Energy loss PDFs are plotted for various thicknesses of water absorbers, where the thickness is expressed in units of mean free path (mfp). For visual clarity, the energy-loss PDFs have been scaled on both the abscissa and ordinate. The single event energy loss is expressed as a fraction of the mean energy lost in the entire absorber thickness, or $(\Delta-\Delta_{av})/\Delta_{av}$.

Next, we calculate the proton range using an analytical approach, which may be faster and more practical than the numerical approach for many clinical calculations. We begin by noting that the interval of proton range of interest typically extends from 1mm (about the size of a voxel in an anatomic image of patient anatomy) to about 30cm (about the midline of a large adult male's pelvis, the deepest site in the human body). These ranges correspond to 11MeV and 220MeV, respectively, as seen in figure. More importantly, Above figure reveals that the relationship between the logarithm of range and logarithm of energy is almost linear. This is fortuitous because this means that the range follows a very simple power law, as realized by Bragg and Kleeman and others early in the last century. Thus, the range of a proton may be calculated using the Bragg–Kleemann rule, or

$$R(E)=\alpha E^{p},$$

where, as before, α is a material-dependent constant, E is the initial energy of the proton beam, and the exponent p takes into account the dependence of the proton's energy or velocity.

The uncertainty in proton range depends on many factors. For example, the uncertainty in a range measurement depends on the precision and accuracy of the measurement apparatus and, in some cases, on the experimenter's skill. A common concern in clinical proton therapy is the uncertainty in calculated range, e.g. in calculating the settings of the treatment machine for a patient's treatment. The uncertainty in range may depend on the knowledge of the proton beam's energy distribution and on properties of all range absorbing materials in the beam's path. These properties include elemental composition, mass density, and linear stopping power. The linear stopping powers deduced from computed tomography scans have many additional sources of uncertainty, including imperfections in the calibration of CT scanners (i.e. the method used to convert image data from Hounsfield Units to linear stopping power), partial volume effects, motion artifacts, and streaks artifacts.

Energy and Range Straggling

We approximated the energy loss rate by assuming that the slowing of ions occurs in a smooth and continuous manner. In fact, we considered the mean energy loss rate and neglected variations in the energy loss rates of individual protons. For many clinical calculations, these assumptions are valid and lead to a reasonably good first-order approximation. However, the accumulation of many small variations in energy loss, termed energy straggling or range straggling, is one of the physical processes that strongly governs the shape of a proton Bragg curve. Thus, an understanding of range straggling is key to understanding the characteristics of proton dose distributions.

Figure plots the relative energy loss probability density functions (PDFs) for protons transmitted through water absorbers of various thickness. The curves have been normalized to enhance visual clarity. Apparently, thick absorbers result in a symmetric distribution of energy losses, whereas thin absorbers yield curves that are asymmetric with modes less than the mean and long tails of large energy losses. In principle, straggling PDFs may be calculated numerically from first principles, but usually theoretical approaches are used. Later in the section, three of the most commonly used straggling theories are described.

Having examined the mean energy loss rate in modest detail, and having conceptually introduced energy straggling, it is instructive to understand how these effects are related mathematically before delving into straggling theory. To understand these relationships, let us consider the moments of the ion energy PDF, or

$$M_n = \int_0^{\Delta_1^{max}} \Delta^n f(\Delta)\, d\Delta$$

where Δ is the energy loss of an ion in traversing an absorber, $f(\Delta)d\Delta$ is the probability of energy loss occurring in the interval from Δ *to* $\Delta + d\Delta$, n is the order of the moment, and Δ_1^{max} is from equation $\varphi_L(\lambda_L) = \frac{1}{\pi}\int_0^{\infty} e^{-y(Iny-\lambda_L)} \sin(\pi y)dy$. The zeroth moment is the total collision cross section, the first moment is the (mean) electronic stopping power, the second moment corresponds to the width (variance) of the energy straggling distribution, the third moment to its asymmetry (skewness), and the fourth to its kurtosis. The variance, sometimes denoted by σ_Δ^2, or second moment of the straggling distribution $f(\Delta)$, is

$$M_2(\Delta) = \sigma_\Delta^2 = \bar{v} \int_0^{\Delta_1^{max}} \Delta^2 f_1(\Delta) d\Delta,$$

where $\bar{v}$ is the average number of primary collisions per proton traversal.

Next we examine theories for calculating energy straggling proposed by Bohr, Landau, and Vavilov. These theories are valid for thick, intermediate, and thin absorbers, respectively. The criterion for selecting a valid theory for a given absorber thickness is based on a reduced energy loss parameter,

$$k = \frac{\xi}{\Delta_1^{max}}$$

where ξ is the approximate mean energy loss and $\Delta_1^{ü}$ is the maximum energy loss possible in a single event.

According to Bohr's theory, the energy straggling distribution behaves according to a Gaussian PDF, or

$$f(\Delta)d\Delta = \frac{1}{\sigma_\Delta \sqrt{2\pi}} exp \frac{-(\Delta - \bar{\Delta})^2}{2\sigma_\Delta^2},$$

where for non-relativistic ions the variance is given by

$$\sigma_\Delta^2 = 2\pi r_e^2 m_e c^2 z^2 \frac{NZ}{\beta^2} \Delta_1^{max} \rho x,$$

where p is the mass density and x is the thickness of the absorber. Bohr's theory assumes that the absorber is thick enough that there are many individual collisions (i.e. the central limit theorem holds), that the ion velocity does not decrease much in crossing the absorber, and that the absorber is made of unbound electrons. For most applications in proton therapy, Bohr's theory provides adequate accuracy. Several authors have reported convenient power law approximations to estimate sigma as a function of the proton beam range or

$$\sigma_\Delta \approx kR_0^m,$$

where R_0 is the range in water in centimeters for a mono-energetic proton beam, k is a material-dependent constant of proportionality, and the exponent is empirically determined. For protons in water, k = 0.012 and m = 0.935.

Landau's theory relaxed the assumption that the central limit holds, i.e. there are relatively fewer individual collisions in an intermediate thickness absorber, and used an approximate expression for Δ_1. In this case, we have

$$f(\Delta, \rho x)d\Delta = \frac{1}{\xi}\varphi_L(\lambda_L),$$

where the parameter $\varphi_L(\lambda_L)$ roughly corresponds to the deviation from the mean energy loss and was defined by Landau as

$$\varphi_L(\lambda_L) = \frac{1}{\pi}\int_0^\infty e^{-y(Iny-\lambda_L)}\sin(\pi y)dy.$$

Evaluation of the integral in equation $\varphi_L(\lambda_L) = \frac{1}{\pi}\int_0^\infty e^{-y(Iny-\lambda_L)}\sin(\pi y)dy$ is straightforward.

Vavilov's theory is in essence a generalization of Landau's theory that utilizes the correct expression for Δ_1 and is given by

$$f(\Delta, \rho\xi)d\Delta = \frac{1}{\xi}\varphi_v(\lambda_v, k, \beta^2)d\lambda_v,$$

where

$$\varphi_v(\lambda_v, k, \beta^2) = \frac{1}{\pi}e^{k(1+\beta^2\gamma)}\int_0^\infty e^{kf_i}\cos(\lambda_v y + kf_2)dy,$$

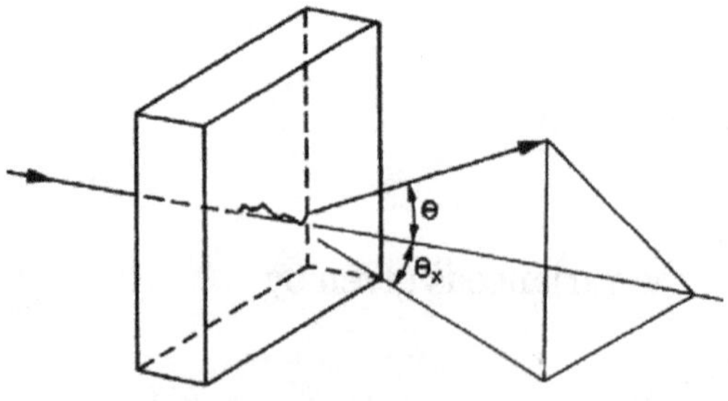

Figure: Schematic diagram showing the trajectory of a proton undergoing multiple Coulomb scattering events. θ is the root mean square (rms) space scattering angle and θ_x is the projected rms scattering angle (reproduced with permission from Leo).

And

$$\lambda_v = \frac{\Delta - \bar{\Delta}}{\Delta_1^{max}} - k(1+\beta^2-\gamma),$$

where γ is Euler's constant. The evaluation of Vavilov's theory is computationally more expensive than that of Bohr's or Landau's.

Multiple Coulomb Scattering

As illustrated in figure, a proton passing close to the nucleus will be elastically scattered or deflected by the repulsive force from the positive charge of the nucleus. While the proton loses a negligible amount of energy in this type of scattering, even a small change in its trajectory can be of paramount importance. In fact, it is necessary to take into account Coulomb scattering when designing beam lines and treatment heads and in calculations of dose distributions in phantoms or patients, e.g. with treatment planning systems.

To characterize the amount that a beam is deflected by scattering, we use the quantity of scattering power, which is defined as

$$T = \mathrm{d} < \theta^2 > / dx$$

where $< \theta^2 >$ is the mean squared scattering angle and x is the thickness of absorber through which the proton has traveled. Similarly, the mass scattering power is simply T / ρ, where ρ is the mass density of the absorber material. Notice that the definition of scattering power utilizes the mean square of the scattering angle; scattering is symmetric about the central axis, and therefore the mean scattering angle is zero and contains no useful information. Also, note that the value of the scattering power depends on the material properties and dimensions of the absorber being considered.

Predictions of elastic Coulomb scattering are commonly classified according to the number of individual scattering events (N_s) that occur in a given absorber. For single scattering $(N_s = 1)$, Rutherford scattering theory applies. Plural scattering $(1 < N_s < 20)$ is difficult to model theoretically and is not discussed further here. For multiple Coulomb scattering $(MCS; N_s \geq 20)$, the combined effect of all Ns scattering events, may be modeled by using a statistical approach to predict the probability for a proton to scatter by a net angle of deflection, commonly denoted by θ (above figure).

It is instructive to briefly examine Rutherford's theory of single scattering. The differential cross section $d\sigma$ for scattering into the solid angle $d\Omega$ is given by

$$\frac{d\sigma}{d\Omega} = z_1^2 z_2^2 r_e^2 \frac{(m_e c / \beta p)^2}{4 \sin^4(\theta / 2)}$$

where z_1 is the charge of the projectile, z_2 is the atomic number of the absorber material, r_e is the classical electron radius, m_e is the mass of the electron, c is the speed of light $\beta = v / c$, p is the ion momentum, and θ is the scattering angle of the proton. The angular dependence is governed by the term $1 / sin^4(\theta / 2)$, i.e. in individual scattering events, the proton is preferentially scattered in the forward direction, at very small angles.

In clinical proton therapy, most objects of interest are thick enough to produce a great many scattering events. Thus, for clinical proton therapy, we are usually more interested in the net effect that many small-angle scattering events have on many protons, e.g. how beam line scattering in the treatment head influences the spatial distribution of dose in a patient.

Rigorous theoretical calculations of MCS are quite complex. The most complete theory was proposed by Molière. Assuming that scattered particles are emitted at small deflection angles (i.e. the small-angle approximation in which $\sin(\theta) \approx \theta$),

$$P(\theta)\, d(\Omega) = \eta d\eta \left(2\exp(-\eta^2) + \frac{F_1(\eta)}{B} + \frac{F_2(\eta)}{B^2} \ldots \right),$$

where $\eta = \theta / (\theta_1\, B^{1/2})$, $\theta_1 = 0.3965\ (zZ / p\beta)\ \sqrt{\rho\delta x / A}$ and $d(\Omega)$ is the solid angle into which the particles are scattered. The functions $F_k(\eta)$ are defined as

$$F_k(\eta) = \frac{1}{k!} \int J_0(\eta y)\ \exp\left(\frac{-y^2}{4}\right) \left[\frac{y^2}{2} \ln\left(\frac{y^2}{4}\right)\right]^k y dy,$$

where $J_o(\eta\gamma)$ is a Bessel function. Values of F_i have been tabulated in the literature by several authors. Parameter B in equation $P(\theta)\, d(\Omega) = \eta d\eta \left(2\exp(-\eta^2) + \frac{F_1(\eta)}{B} + \frac{F_2(\eta)}{B^2} \ldots \right)$, may be found by numerical methods, by solving

$$g\ (B) = \ln B - B + \ln \gamma - 0.154 = 0,$$

where $\gamma = 8831\, q\, z^2\ \rho\delta x / (\beta^2 A\Delta)$ and $\Delta = 1.13 + 3.76\ (Zz / 137\beta)^2$. The following are symbols representing properties of the absorber: Z is the atomic number, A is the atomic mass, δx is the thickness, and ρ is the mass density. The proton momentum is denoted by p, $\beta = v / c$, and $z = 1$ is the proton charge. Even though we have not presented all of the details, clearly the evaluation of Molière's theory is indeed complex. Consequently, considerable attention has been paid in the literature to developing simpler formulae; the simplicity usually comes at the cost of reduced accuracy in modeling scattering at large angles. Gottschalk discussed several of these methods in detail. One approximation method takes the form of a Gaussian distribution,

$$P(\theta) \approx \frac{2\theta}{<\theta^2>} \exp\left(\frac{-\theta^2}{<\theta^2>}\right) d\theta,$$

where $<\theta^2>^{1/2}$ is the root mean square (rms) scattering angle or the width parameter of the Gaussian distribution.

Gottschalk proposed a differential approximation to Molière's theory to predict the scattering power according to

$$T_{dM} = f_{dM}\ (pv,\, p_1 v_1) \times \left(\frac{E_s}{pv}\right)^2 \frac{1}{X_s},$$

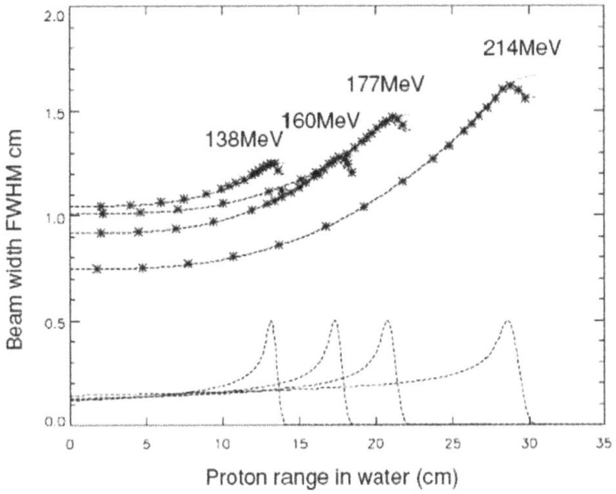

Figure : Broadening of the beam width in water due to multiple Coulomb scattering.

where E_S = 15MeV, p is the momentum and v is the velocity of the proton, p_1 and v_1 are the initial momentum and velocity, X_S is scattering length of the material, and f_{dM} is a material in dependent nonlocal correction factor given by

$$f_{dM} = 0.5244 + 10.1975\lg\left(1-(pv/p_1v_1)^2\right) +$$
$$0.2320lg(pv)\ 0.0098\lg(pv)\lg\left(1-(pv/p_1v_1)^2\right).$$

The factor f_{dM} takes into account the accumulation of scattering that occurs as the proton slows from v_1 to v and is material independent. The scattering length is given by

$$\frac{1}{\rho X_s} = \alpha N \rho r_e^2 \frac{Z^2}{A}(2\log(33219(AZ)^{-1/3} - 1),$$

where α is the fine structure constant, N is Avogadro's number, r_e is electron radius, and Z, A, and ρ are the atomic number, atomic weight, and density of the target material.

It is sometimes convenient to project the expected scattering angle $<\theta>$ in 3D space to the corresponding value in a plane, denoted $<\theta_x>$, according to

$$<\theta_x>^2 = <\theta_x^2>/2.$$

Also, in proton therapy the lateral displacement (r) of a proton beam is typically calculated from the scattering angle. Under the Gaussian approximation, we have

$$P(r)\,dr = 6r\,(<\theta^2>t^2)^{-1}\exp(-3r^2/(<\theta^2>t^2)\,dr,$$

where $t = x/L_{rad}$, and L_{rad} is the radiation length of the material, which can be calculated easily or looked up in tables. The mean square lateral displacement is given by

$$<r^2> = <\theta^2>t^2/3.$$

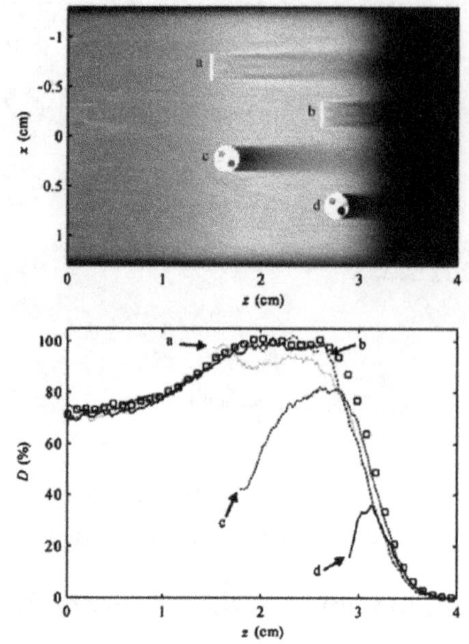

Small implanted fiducial markers can create clinically significant dosimetric cold spots in proton therapy beams. (Upper) 2D dose distribution as a function of depth in water (z) and crossfield position (x) from a Monte Carlo simulation of rangemodulated proton beam incident on a water phantom containing tantalum localization markers oriented (a, b) parallel to the beam axis and (c, d) perpendicular to the beam axis. The range and modulation width are typical for uveal melanoma treatments. (Lower) simulated absorbed dose (D) as a function of depth (z) in the water phantom at various off-axis positions. The perturbed depth dose profiles are parallel to the beam axis and pass through the center of markers a–d in the plot above. For visual clarity, portions of the perturbed dose profiles upstream of the markers are not shown. An unperturbed beam is plotted with open squares.

A power law approximation for rms displacement of protons is

$$r_{rms} = aR^b ,$$

where a is a unitless material constant, R is the water-equivalent proton beam range in cm, and the exponent b governs the range dependence. For protons in water, a = 0.0294 and b = 0.896.

Interaction of Neutron with Matter

Since the neutron is electrically neutral, it interacts only weakly with matter into which it can penetrate deeply. Contrary to X-rays, which interact dominantly with the electron shell of the atom, the neutron does on the level of the nucleus. Therefore the neutron is quite sensitive to light atoms like hydrogen, oxygen, etc. which have much higher interaction probability with neutrons than with X-rays. In contrast to this, metals comparatively show lower interaction probability with neutrons, thus allowing quite high penetration depth. Elements of similar atomic number Z (= number of protons) likewise differentiate easier with neutrons than with X-rays. A comparison of neutron and X-ray attenuation is given in the section Comparison with X-ray. For neutron transmission the universal law of attenuation of radiation passing through matter (Beer-Lambert law) is valid.

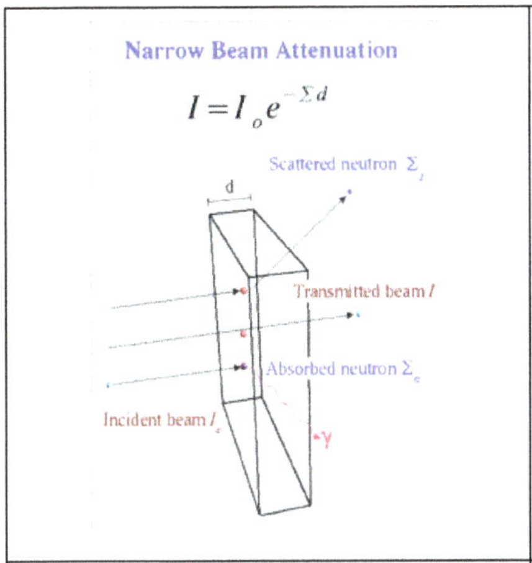

Figure Neutron scattering and capture interaction probabilities.

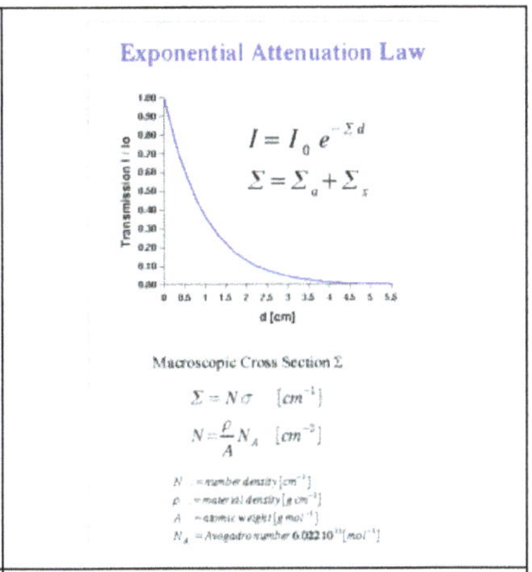

Figure Exponential attenuation of neutrons in matter: Beer-Lambert law.

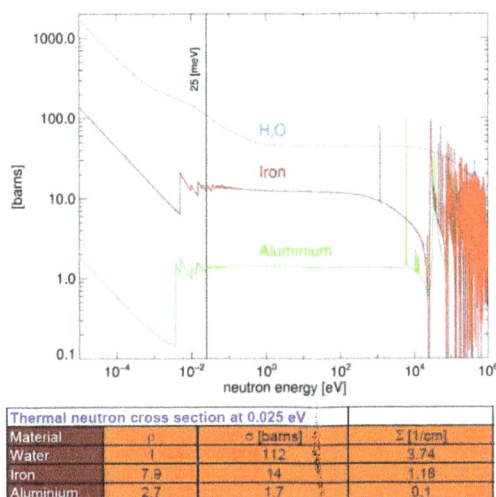

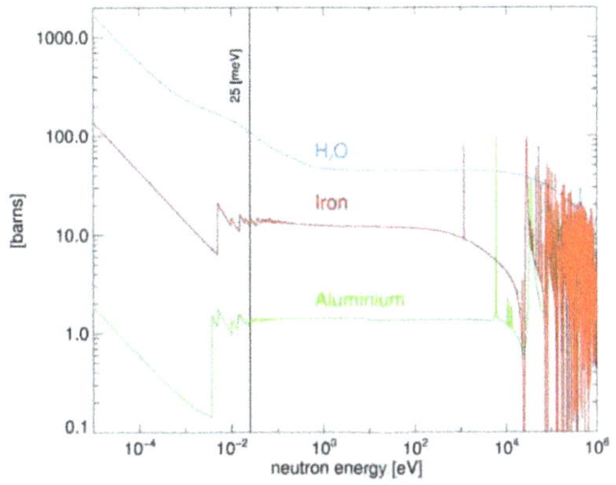

Thermal neutron cross section at 0.025 eV			
Material	ρ	σ [barns]	Σ [1/cm]
Water	1	112	3.74
Iron	7.9	14	1.18
Aluminium	2.7	1.7	0.1

Figure Energy dependence of neutron cross sections shown for aluminium, iron and the water molecule

Neutron matter interaction probability is high at thermal or cold energies. It is given as microscopic cross-section data σ in units of barns i.e. 10^{-24} cm^2. Neutrons for transmission radiography are therefore extracted from a moderator (e.g. tank containing heavy water), which slows down neutrons produced in a spallation source or a research reactor in the MeV energy range.

The transmission behavior of a mono-energetic narrow neutron beam can be described by the basic law of radiation attenuation in matter. The ratio between the emerging neutron flux I and incident flux I_0 is called transmission T. Quantitative data about the material composition (e.g. hydrogen content), can be derived from neutron transmission measurements in the case of known sample shape and dimension d. Hereby the relation defining the macroscopic neutron interaction cross section and the evaluated neutron cross section data are taken from a database. The simple exponential attenuation law does not hold for all situations. Thick samples or strongly scattering (e.g. hydrogen) or absorbing materials (e.g. containing strong absorbers like boron, gadolinium) show a deviation, due to multiple scattering effects or the need to take the changing neutron energy spectrum into account.

Interaction of Heavy Charged Particles with Matter

Nature of Interaction of Charged Particles with Matter

Since the electromagnetic interaction extends over some distance, it is not necessary for the light or heavy charged particle to make a direct collision with an atom. They can transfer energy simply by passing close by. Heavy charged particles, such as fission fragments or alpha particles interact with matter primarily through coulomb forces between their positive charge and the negative charge of the electrons from atomic orbitals. On the other hand, the internal energy of an atom is quantised, therefore only certain amount of energy can be transferred. In general, charged particles transfer energy mostly by:

- Excitation. The charged particle can transfer energy to the atom, raising electrons to a higher energy levels.
- Ionization. Ionization can occur, when the charged particle have enough energy to remove an electron. This results in a creation of ion pairs in surrounding matter.

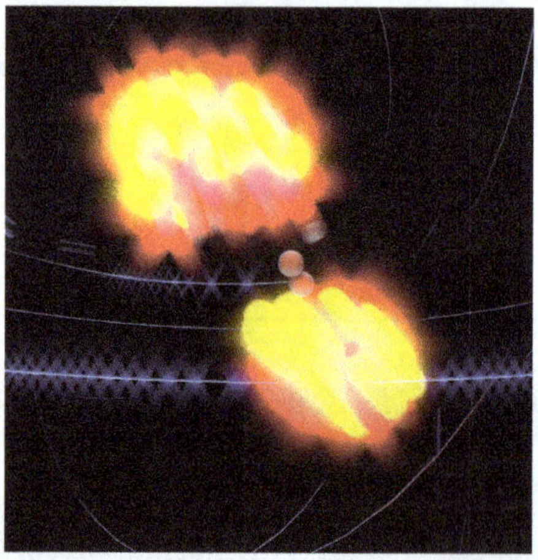

Fission fragments after a nucleus fission. Fission fragments interact strongly with the surrounding atoms or molecules traveling at high speed, causing them to ionize.

Creation of pairs requires energy, which is lost from the kinetic energy of the charged particle causing it to decelerate. The positive ions and free electrons created by the passage of the charged particle will then reunite, releasing energy in the form of heat (e.g. vibrational energy or rotational energy of atoms). This is the principle how fission fragments heat up fuel in the reactor core. There are considerable differences in the ways of energy loss and scattering between the passage of light charged particles such as positrons and electrons and heavy charged particles such as fission fragments, alpha particles, muons. Most of these differences are based on the different dynamics of the collision process. In general, when a heavy particle collides with a much lighter particle (electrons in the atomic orbitals), the laws of energy and momentum conservation predict that only a small fraction of the massive particle's energy can be transferred to the less massive particle. The actual amount of transferred energy depends on how closely the charged particles passes through the atom and it depends also on restrictions from quantisation of energy levels.

The distance required to bring the particle to rest is referred to as its range. The range of fission fragments in solids amounts to only a few microns, and thus most of the energy of fission is converted to heat very closeto the point of fission. In case of gases the range increases to a few centimeters in dependence of gas parameters (density, type of gas etc.) The trajectory of heavy charged particles are not greatly affected, because they interacts with light atomic electrons. Other charged particles, such as the alpha particles behave similarly with one exception – for lighter charged particles the ranges are somewhat longer.

Stopping Power – Bethe Formula

A convenient variable that describes the ionization properties of surrounding medium is the stopping power. The linear stopping power of material is defined as the ratio of the differential energy loss for the particle within the material to the corresponding differential path length:

$$S(T) = -\frac{dT}{dx} = n_{ion}\,\bar{I},$$

where T is the kinetic energy of the charged particle, n_{ion} is the number of electron-ion pairs formed per unit path length, and I denotes the average energy needed to ionize an atom in the medium. For charged particles, S increases as the particle velocity decreases. The classical expression that describes the specific energy loss is known as the Bethe formula. The non-relativistic formula was found by Hans Bethe in 1930. The relativistic version was found also by Hans Bethe in 1932.

$$S(T) = \frac{4\pi Q^2 e^2 n Z}{m\beta^2 c^2}\left[In\left(\frac{2mc^2\gamma^2\beta^2}{\bar{I}}\right) - \beta^2\right]$$

In this expression, m is the rest mass of the electron, β equals to v/c, what expresses the particle's velocity relative to the speed of light, γ is the Lorentz factor of the particle, Q equals to its charge, Z is the atomic number of the medium and n is the atoms density in the volume. For nonrelativistic particles (heavy charged particles are mostly nonrelativistic), dT/dx is dependent on $1/v^2$. This is can be explained by the greater time the charged particle spends in the negative field of the electron, when the velocity is low.

The stopping power of most materials is very high for heavy charged particles and these particles have very short ranges. For example, the range of a 5 MeV alpha particle is approximately only 0,002 cm in aluminium alloy. Most alpha particles can be stopped by an ordinary sheet of paper or living tissue. Therefore the shielding of alpha particles does not pose a difficult problem, but on the other hand alpha radioactive nuclides can lead to serious health hazards when they are ingested or inhaled (internal contamination).

Specifics of Fission Fragments

The fission fragments three two key features (somewhat different from alpha particles or protons), which influence their energy loss during its travel through matter.

- High initial energy. Results in a large effective charge.
- Large effective charge. The fission fragments start out with lack of many electrons, therefore their specific loss is greater than alpha's specific loss, for example.
- Immediate electron pickup. Results in changes of (-dE/dx) during the travel.

These features results in the continuous decrease in the effective charge carried by the fission fragment as the fragment comes to rest and continuous decrease in -dE/dx. The resulting decrease in -dE/dx (from the electron pickup) is larger than the increase that accompanies a reduction in velocity. The range of typical fission fragment can be approximately half that of a 5 MeV alpha particle.

Bragg Curve

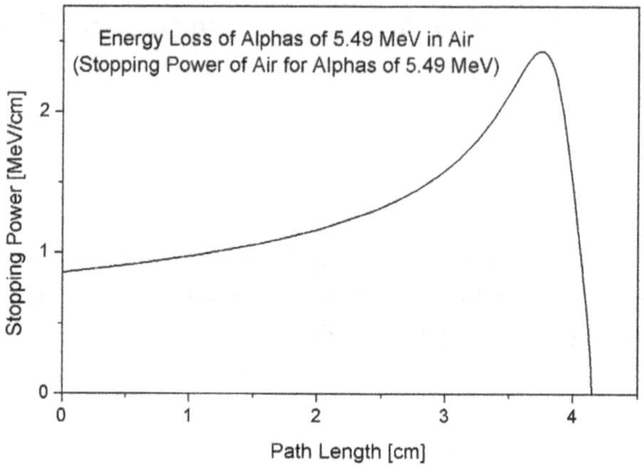

Bragg Curve is typical for heavy charged particles and plots the energy loss during its travel through matter.

The Bragg curve is typical for heavy charged particles and describes energy loss of ionizing radiation during travel through matter. For this curve is typical the Bragg peak, which is the result of $1/v^2$ dependency of the stopping power. This peak occurs because the cross section of interaction increases immediately before the particle come to rest. For most of the track, the charge remains unchanged and the specific energy loss increases according to the $1/v^2$. Near the end of the track, the charge can be reduced through electron pickup and the curve can fall off.

The Bragg curve also differs somewhat due to the effect of straggling. For a given material the range will be the nearly the same for all particles of the same kind with the same initial energy. Because the details of the microscopic interactions undergone by any specific particle vary randomly, a small variation in the range can be observed. This variation is called straggling and it is caused by the statistical nature of the energy loss process which consists of a large number of individual collisions.

This phenomenon, which is described by the Bragg curve, is exploited in particle therapy of cancer, because this allows to concentrate the stopping energy on the tumor while minimizing the effect on the surrounding healthy tissue.

References

- Nuclear-Forces: scholarpedia.org, Retrieved 27 March 2018
- What-is-elastic-scattering: wisegeek.com, Retrieved 05 July 2018
- Nuclear-force, modern-physics: physics.tutorvista.com, Retrieved 25 May 2018
- Weak-force-49254: livescience.com, Retrieved 12 April 2018
- Neutron-interaction-with-matter: psi.ch, Retrieved 06 April 2018

5

Nuclear Radiation Detectors

A radiation detector is an instrument that is used for tracking, detecting or identifying particles produced by nuclear decay, particle accelerator reactions or cosmic radiation. These devices can measure particle characteristics of energy, momentum, spin, charge, etc. This chapter strives to provide an overview of nuclear radiation detectors, particularly with reference to the bubble chamber, cloud chamber, gas detector, Cherenkov detectors, scintillation counters, etc.

Bubble Chamber

Bubble chamber is a device used for detecting charged particles and other radiation by means of tracks of bubbles left in a chamber filled with liquid hydrogen or other liquefied gas.

The bubble chamber was invented in 1952 by the American scientist D. Glaser. The superheated liquid can exist from some time τ, after which it starts toboil. If during the time interval τ an ionizing particle enters the chamber, its path will be marked by a line of vapor bubbles that can

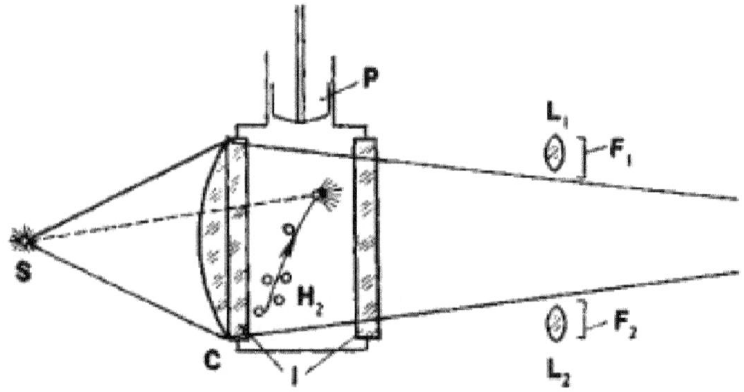

Figure: Diagram of a bubble chamber using liquid hydrogen (H_2).Expansion is effected through the piston (P). Light from impulsesource (S) passes through the glass illuminator (I) and condenser(C). After being dispersed by the bubbles, the light is fixed on films(F_1 and F_2) by means of lenses (L_1 and L_2).

be photographed. The bubble chamber may be thought ofas a Wilson chamber "in reverse"; that is, instead of drops of liquid in a supersaturated vapor there are vapor bubbles in a superheated liquid. This analogy, however, is purely superficial, since the mechanisms for the formation of drops in the Wilson chamber and of bubbles in the bubblechamber are quite different.

The action of the bubble chamber is due to the formation of centers of boiling—nucleus bubbles in the path of aparticle and to their growth to dimensions exceeding the critical value:

$$r_{cr} = 2\sigma\left[p_0 \exp\left(-\frac{V}{V'}\frac{p_0 - p_{cr}}{p_0}\right) - p\right]^{-1}$$

Here, r_{cr} is the radius of the critical bubble, σ is the surface tension of the liquid, p_0 is the saturated vapor pressure, p_{cr} is the critical pressure, p is the vapor pressure in the superheated liquid, V is the specific volume of the liquid, and V′ is the specific volume of the vapor. For the formation of a bubble larger than the critical bubble, energy of the order of a few hundred electron volts must be released in a volume of radius ~ 10^{-6} cm in a time ~ 10^{-6} sec. This energy is released in the deceleration of the electrons, here delta electrons, that are ejected from the atoms of the liquid by the particle being detected. The time required for the bubbles to grow to the dimensions suitable for photography (0.1 - 0.3 mm) ranges from a few milliseconds to tens of milliseconds for different bubble chambers.

Liquid hydrogen and deuterium are preferred for cryogenic bubble chambers, and propane (C_3H_8), various freons, xenon, and a xenon-propane mixture are used in heavy-liquid bubble chambers.

The liquid in a bubble chamber is superheated by rapidly lowering the pressure from the initial value $p_i > p_0$ to $p < p_0$. The pressure drop is accomplished in 5–15 m sec either by displacing a piston, in the case of liquid hydrogen chambers (figure), or by discharging external pressure from a cavity bounded by a rubber diaphragm, in the case of heavy-liquid chambers.

The particles are admitted to the bubble chamber when itis at maximum sensitivity. After the time necessary for thebubbles to become sufficiently large, the chamber isilluminated and the tracks are detected through thetechnique of stereoscopic photography with two to fourlenses. After photography, the pressure is raised to theprevious level, the bubbles disappear, and the bubblechamber once again is ready for operation. The entireoperating cycle of the chamber runs less than 1 sec, andthe sensitive time period is ~10—40 msec.

Bubble chambers, except those containing xenon, areplaced in strong magnetic fields. This allows adetermination of the momentum of a charged particle bymeasuring the radius of curvature ρ of the particle's trajectory:

$$kc = 300H\rho / \cos\phi$$

Here, ϕ is the angle between the direction of the magneticfield H and the momentum k of the particle and c is thespeed of light. Track distortions in the bubble chamber aresmall and are due primarily to the multiple scattering ofparticles. With the use of precision measuring equipment itis possible to determine the spatial positions of the tracksand their curvatures with a greater degree of accuracy.

Bubble chambers generally are used to detect either interaction events of high energy particles with the nuclei of the chamber liquid or the decay events of particles. In the former case, the chamber liquid acts as both a detecting medium and target medium in below figure. The size of the chamber determines its effectiveness in detecting the various processes of interaction or decay. Neutral particles, such as gamma rays and neutrons, are detected on the basis of their interaction with the chamber liquid (table). Bubble chambers with a volume of a few hundred liters are the most common, but there are much larger.

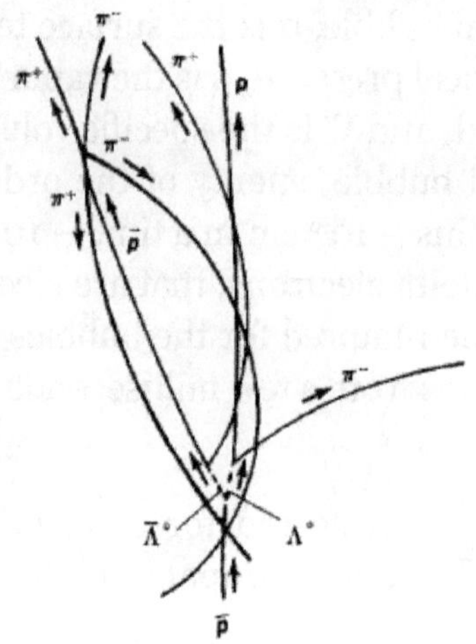

Figure: Detection in a liquid-hydrogen chamber of the nuclear reaction

$$\overline{p}+p\rightarrow\overline{A}^{\circ}+A^{\circ}\rightarrow\pi^{-}+p$$
$$\quad\quad\quad\quad\quad\quad\llcorner\ \pi^{+}+p$$

The antiproton ρ^{-} formed upon the decay of the anti-lambda hyperon $\overline{\wedge}^{\circ}$ collides with proton ρ and is annihilatedas a result of the reaction:

$$\rho^{-}+\rho\rightarrow 2_{\pi^{+}}+2_{\pi^{-}}$$

Here $\wedge^{\circ}$ is a lambda hyperon and π and π^{+} are pions.

The Mirabel' hydrogen chamber, for example, inthe accelerator of the Institute of Highenergy Physics ofthe Academy of Sciences of the USSR, has a volume of 10m^3; the hydrogen chamber in the accelerator of the FermiNational Accelerator Laboratory in the US has a volume of 25 m^3.

The main advantages of bubble chambers are an isotropic spatial sensitivity to particle registration and a high precision in the measurement of particle momenta. One drawback is a low degree of control over the selection of necessary events of particle interaction or decay.

Table: Characteristics of liquids used most often In bubble chambers

Liquid	*Operating conditions*			*Probabilityofrecordinga 500-MeV$_\gamma$quantumover alength of50 cm*	*Probabilityofrecordin-ga 1-GeVneutronover alength of50 cm*
	Pres-sure(atm)	*Tempera-ture(°C)*	*Density (g/cm³)*		
Hydrogen	4.7	−246	0.07	0.046	0.1
Deuterium	5.2	−240	0.13	0.055	0.185
Helium	0.3	−270	0.124	0.053	0.113
Propane	21	58	0.44	0.36	0.340
Xenon	26	−19	2.2	1	

Cloud Chamber

A unique device for detection and measurement is the Cloud Chamber, invented by the British physicist Charles Wilson in 1911. The Cloud chamber makes it possible to visually see the path of ionizing radiation thus making it possible to photograph it. The cloud chamber consists of a plastic or glass container, which sits on dry ice. A dark cloth is saturated with alcohol and placed around the inside of the container near the top. A small radioactive material may be suspended from the lid of the container. In the chamber, the alcohol evaporates from the cloth and condenses as it reaches the cold region created by the dry ice at the floor of the container. Just above the floor of the chamber there is a region where the alcohol vapor does not condense unless there are seeds around, so that drops of alcohol can form. This condition is similar to that of seeding clouds with a chemical to form rain. The idea is that only seeds available in the chamber are those of ions produced by the interaction with radiation. The resulting trail of alcohol droplets can be seen against the black background in the bottom of the chamber.

These are only a few of the devices commonly utilized for purposes of detection and measurement of radioactivity and radiation. It is important to understand that when working with radioactivity/radiation, due to our inability to sense radiation, we need them to assist us in detecting the presence of radiation and we also need them to help monitor the radiation.

Structure and Operation

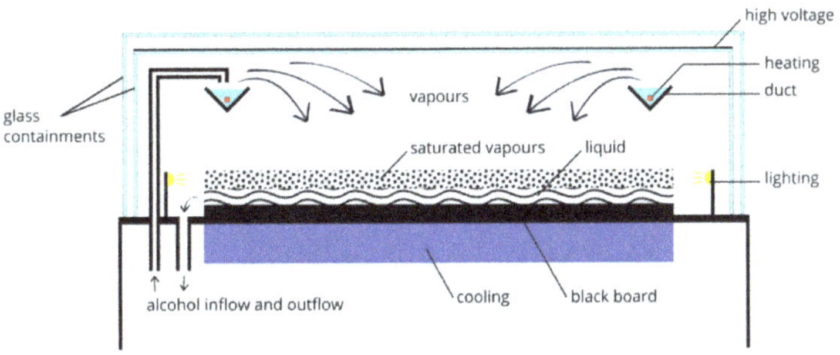

Figure: A diffusion-type cloud chamber. Alcohol (typically isopropanol) is evaporated by a heater in a duct in the upper part of the chamber. Cooling vapor descends to the black refrigerated plate, where it condenses. Due to the temperature gradient a layer of supersaturated vapor is formed above the bottom plate. In this region, radiation particles induce condensation and create cloud tracks.

Diffusion-type cloud chambers will be discussed here. A simple cloud chamber consists of the sealed environment, a warm top plate and a cold bottom plate. It requires a source of liquid alcohol at the warm side of the chamber where the liquid evaporates, forming a vapor that cools as it falls through the gas and condenses on the cold bottom plate. Some sort of ionizing radiation is needed.

Methanol, isopropanol, or other alcohol vapor saturates the chamber. The alcohol falls as it cools down and the cold condenser provides a steep temperature gradient. The result is a supersaturated environment. As energetic charged particles pass through the gas they leave ionization trails. The alcohol vapor condenses around gaseous ion trails left behind by the ionizing particles. This occurs because alcohol and water molecules are polar, resulting in a net attractive force toward a nearby free charge. The result is a misty cloud-like formation, seen by the presence of droplets falling down to the condenser. When the tracks are emitted radially outward from a source, their point of origin can easily be determined.

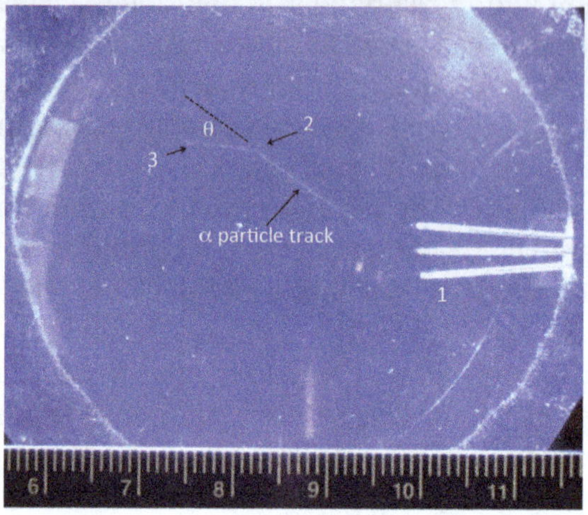

Figure: In a diffusion cloud chamber, a 5.3 MeV alpha-particle track from a Pb-210 pin source near Point (1) undergoes Rutherford scattering near Point (2), deflecting by angle theta of about 30 degrees. It scatters once again near Point (3). The target nucleus in the chamber gas could have been a nitrogen, oxygen, carbon, or hydrogen nucleus. It received enough kinetic energy in the elastic collision to cause a short visible recoiling track near Point.

Just above the cold condenser plate there is a volume of the chamber which is sensitive to ionization tracks. The ion trail left by the radioactive particles provides an optimal trigger for condensation and cloud formation. This sensitive volume is increased in height by employing a steep temperature gradient, and stable conditions. A strong electric field is often used to draw cloud tracks down to the sensitive region of the chamber and increase the sensitivity of the chamber. The electric field can also serve to prevent large amounts of background "rain" from obscuring the sensitive region of the chamber, caused by condensation forming above the sensitive volume of the chamber, thereby obscuring tracks by constant precipitation. A black background makes it easier to observe cloud tracks. Typically, a tangential light source is needed. This illuminates the white droplets against the black background. Often the tracks are not apparent until a shallow pool of alcohol is formed at the condenser plate.

If a magnetic field is applied across the cloud chamber, positively and negatively charged particles will curve in opposite directions, according to the Lorentz force law; strong-enough fields are difficult to achieve, however, with small hobbyist setups.

Other Particle Detectors

The bubble chamber was invented by Donald A. Glaser of the United States in 1952, and for this, he was awarded the Nobel Prize in Physics in 1960. The bubble chamber similarly reveals the tracks of subatomic particles, but as trails of bubbles in a superheated liquid, usually liquid hydrogen. Bubble chambers can be made physically larger than cloud chambers, and since they are filled with much-denser liquid material, they reveal the tracks of much more energetic particles. These factors rapidly made the bubble chamber the predominant particle detector for a number of decades, so that cloud chambers were effectively superseded in fundamental research by the start of the 1960s.

A spark chamber is an electrical device that uses a grid of uninsulated electric wires in a chamber, with high voltages applied between the wires. Energetic charged particles cause ionization of the gas along the path of the particle in the same way as in the Wilson cloud chamber, but in this case the ambient electric fields are high enough to precipitate full-scale gas breakdown in the form of sparks at the position of the initial ionization. The presence and location of these sparks is then registered electrically, and the information is stored for later analysis, such as by a digital computer.

Similar condensation effects can be observed as Wilson clouds, also called condensation clouds, at large explosions in humid air and other Prandtl–Glauert singularity effects.

Gas Detector

Gas detectors measure and indicate the concentration of certain gases in an air via different technologies. Typically employed to prevent toxic exposure and fire, gas detectors are often battery operated devices used for safety purposes. They are manufactured as portable or stationary (fixed) units and work by signifying high levels of gases through a series of audible or visible indicators, such as alarms, lights or a combination of signals. While many of the older, standard gas detector units were originally fabricated to detect one gas, modern multifunctional or multi-gas devices are capable of detecting several gases at once. Some detectors may be utilized as individual units to monitor small workspace areas, or units can be combined or linked together to create a protection system.

As detectors measure a specified gas concentration, the sensor response serves as the reference point or scale. When the sensors response surpasses a certain pre-set level, an alarm will activate to warn the user. There are various types of detectors available and the majority serves the same function: to monitor and warn of a dangerous gas level. However, when considering what type of detector to install, it is helpful to consider the different sensor technologies.

Gas Detector Technologies

Gas detectors are categorized by the type of gas they detect: combustible or toxic. Within this broad categorization, they are further defined by the technology they use: catalytic and infrared sensors detect combustible gases and electrochemical and metal oxide semiconductor technologies generally detect toxic gases.

Measurement of Toxic Gases

Electrochemical sensors or cells are most commonly used in the detection of toxic gases like carbon monoxide, chlorine and nitrogen oxides. They function via electrodes signals when a gas is detected. Generally, these types of detectors are highly sensitive and give off warning signals via electrical currents. Various manufacturers produce these detectors with a digital display.

Metal Oxide Semiconductors, or MOS, are also used for detecting toxic gases (commonly carbon monoxide) and work via a gas sensitive film that is composed of tin or tungsten oxides. The sensitive film reacts with gases, triggering the device when toxic levels are present. Generally, metal oxide sensors are considered efficient due their ability to operate in low-humidity ranges. In addition, they are able to detect a range of gases, including combustibles.

Measurement of Combustible Gases

Catalytic sensors represent a large number of gas detector devices that are manufactured today. This technology is used to detect combustible gases such as hydrocarbon, and works via catalytic oxidation. The sensors of this type of detector are typically constructed from a platinum treated wire coil. As a combustible gas comes into contact with the catalytic surface, it is oxidized and the wiring resistance is changed by heat that is released. A bridge circuit is typically used to indicate the resistance change.

Infrared sensors or IR detectors work via a system of transmitters and receivers to detect combustible gases, specifically hydrocarbon vapors. Typically, the transmitters are light sources and receivers are light detectors. If a gas is present in the optical path, it will interfere with the power of the light transmission between the transmitter and receiver. The altered state of light determines if and what type of gas is present.

Common Gas Detector Applications

Although detectors are an essential application for home and commercial safety, they are also employed in numerous industrial industries. Gas detectors are used in welding shops to detect combustibles and toxics and in nuclear plants, to detect combustibles. They are also commonly used to detect hazardous vapors in wastewater treatment plants.

Gas detectors are very efficient in confined spaces where there is no continuous employee occupancy. Such spaces include tanks, pits, vessels and storage bins. Detectors may also be placed at a site to detect toxins prior to occupant entry.

Cherenkov Detectors

Cherenkov radiation is electromagnetic radiation when a charged particle passes through in a dielectric medium at a speed greater than the speed of light in that medium. When a charged particle travels, it disturbs the electric field in the medium and medium becomes electrically polarized. If charged particle has enough speed, a coherent shockwave is emitted. The similar analogy can be though as a sonic boom of a supersonic aircraft. Cherenkov emitting angle can be easily expressed as $\cos(\theta_c) = \frac{1}{n\beta}$ where n is the refractive index of the material and β is v/c. If β is greater than 1/n, Cherenkov radiation will be emitted as shown in figure. Using relativistic kinematics, the threshold energy of Cherenkov radiation can be given by: $E_{threshold}(n, m) = mc^2 \frac{n}{\sqrt{n^2-1}}$. At this energy, the Cherenkov light is emitted along the path of the particle. The same figure right side contains the blueish Cherenkov light observed at the Reed Research Reactor originating from fast electrons. In case of $\Delta n = n - 1 << 1$ approximation, the threshold energy can be rewritten as $E_{threshold}(n, m) = \frac{mc^2}{\sqrt{2\Delta n}}$. Therefore, such a detector could be used to trigger on various particles at various energies for small experiments or for educational purposes, depending on the radiation medium.

Detector Design and Construction

The simplest Cherenkov detector would consist of a barrel with enough volume to produce Cherenkov light and reflective inner surfaces to channel that light into photon counting detectors. In high energy physics such a detector would be a Photo Multiplier Tube (PMT) requiring about 1.5-2 kV for operation. The speed of light in the material to be installed into the barrel has to be smaller the vacuum value (c) to permit Cherenkov radiation. For example water $(v_\gamma = 0.75c)$ would be the a cheap and easily accessible material allowing detection of Cherenkov light for cosmic muons with energy larger than 160 MeV. However using CO2 (Air) would allow detection of cosmic muons of energy threshold of 3.52 (4.4) GeV.

To build the simplest possible detector in the cheapest possible way, possible barrel alternatives have been considered such as a beer keg and a large water dispenser. The simplest solution was found to be a 20L volume plastic trash bin since it permits about 50 cm water depth, large enough to produce Cherenkov light. The inner side of the bin was covered with kitchen grade aluminum foils to improve its reflectivity. A view of the Al covered bin can be seen in below Figure The available PMTs were R7525HA-2 from Hamamatsu with 25mm photocathode diameter and requiring about 1500V for proper operation. Since these PMTs can not operate in water, a simple setup from foam and cardboard was prepared to keep the single PMT above the water level. The signal output from the PMT base can be readout directly with an oscilloscope without needing a pre-amplifier circuit. The light tightness of the finished product was provided by multiple layers of black felt blankets.

Figure: A 20L dust bin with inner surfaces covered with aluminum foils was used as the vessel for Cherenkov light producing material: water.

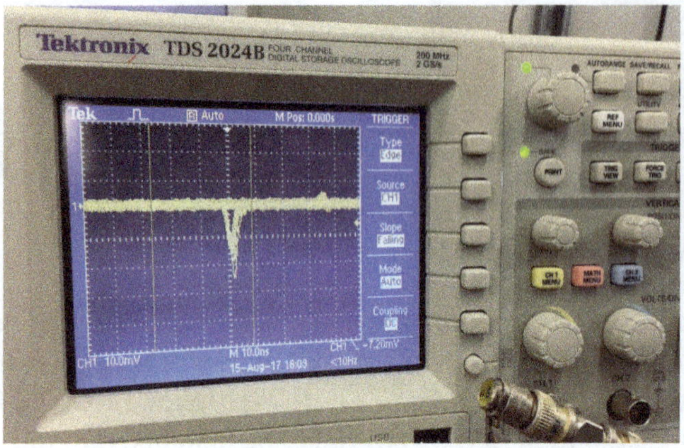

Figure: Background measurements with air in the bin

Detector Tests

The detector was first tested with air inside the bin to have an understanding of the background values. With a threshold as low as -7.2 mV, the background rate is measured to be less than 10Hz. This value is consistent with a background rate originating from the PMT's own dark current and from cosmic rays passing through the PMT window. A screenshot from the oscilloscope in accumulating display mode can be seen in above Figure.

After determination of the rate and pulse hight of the background events, the bin was filled with tap water providing a water height of about 50cm as seen in figure. The signal threshold was set to -30mV eliminating most of the backgrounds. The solid line pulse in the same figure right side acquired after about 10 mins is consistent with a cosmic ray passing through the photocathode, both rate-wise and pulse height-wise which is about -60 mV.

The expected Cherenkov signal on the other hand has to be rarer and should have a much higher pulse hight. Such an event was observed after about 15 mins as shown in figure. Here the pulse hight is about -150 mV as expected from Cherenkov light from a high energy cosmic ray. It is difficult to further elaborate on the cosmic ray's properties without calibrating the detector in a test beam. However, it is thought that by improving the quality of the inner reflecting surfaces, i.e. by collecting more light, the sensitivity of the setup can be increased.

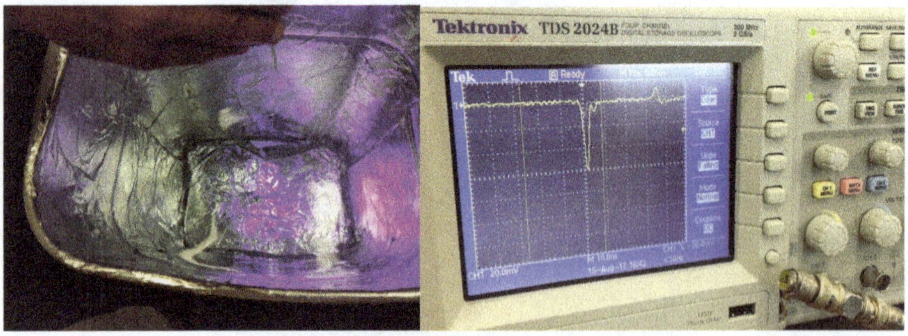

Figure: Background measurements with water in the bin

Of course, adding at least one scintillator to the system and measuring the timing between the scintillator and Cherenkov would largely improve the setup.

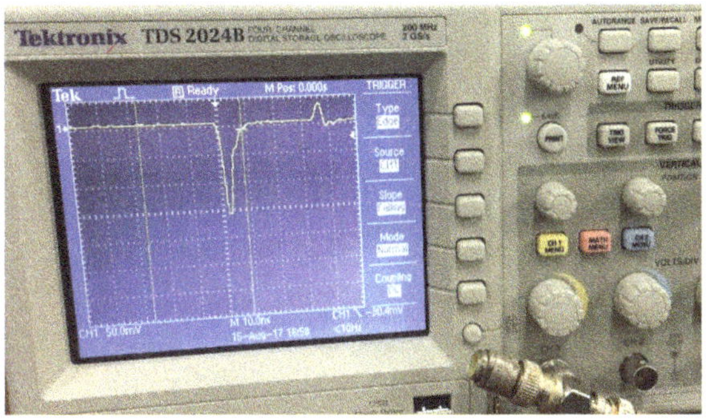

Figure: Cherenkov light measurement with water in the bin

Wire Chamber

Wire chamber is the detectors measure the drift time for the electrons from the ionized gas they contain to the sense wires under the influence of a uniform electric field generated by an applied high voltage. We describe here construction of a "delay" wire chamber (DWC), a simplified version of the well-known multi-wire proportional chamber. The first change is on the cathode: instead of using a conductor plane, it is made of a set of closely distanced wires (about 2mm pitch) placed transversally to the anode wires. When the ionization electrons reach the anode wires, an image charge is produced on the cathode wires. An additional simplification is in the read-out: instead of acquiring data from each individual wire, the signals are accumulated on a delay line which is then read from both ends as shown in below Figure. The timing difference between the two signals (Δt) can be converted into position information by a simple linear approximation: $x = \alpha \times \Delta t + \beta$ where the calibration coefficients α and β are to be determined experimentally. Such a simple detector has somewhat limited position precision, about 200 microns. However given its merits, such as the usage of non-flammable gas (such as Ar-CO_2 mixture) and simplicity of the readout circuitry, such a compromise can be deemed acceptable for small experiments or for educational purposes.

The prototype DWC is prepared for a simple setup: it will be used in the magnetic spectrometer of the SPP proton beamline. Therefore the prototype is even further simplified to measure position only in one plane.

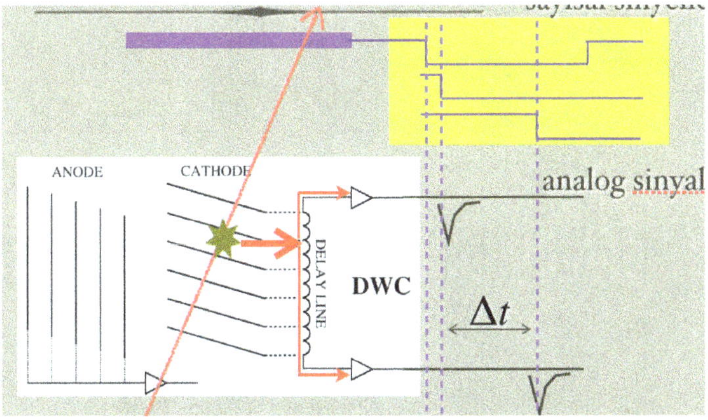

Figure: A typical "delay" wire chamber

DWC Design

The structure to host the anode and cathode wires is made of glass-reinforced epoxy laminate sheets that are also used in the printed circuit board (PBC) making. To cope with the mechanical stress caused by the wires, and to cope with the high voltages necessary for the electron drift, PCB type flame retardant grade 4 (FR4) is selected. FR4 has also the self-extinguishing and near zero water absorption properties, making it ideal for gaseous detector construction. The wire chamber design was made using software tools which are either open source or free for educational use. The mechanical design for the different layers of the DWC prototype is made using openSCAD, available for free for a number of different operating systems. This software tool allows both a 3-D view of the product and also standard engineering drawings. An example to both options can be seen in below Figure. The individual FR4 layers, each having 5mm thickness, are designed with the goal of easy construction and de-construction.

The top and bottom FR4 layers also contain the gas connectors to flush the Ar-CO2 mixture. One could notice the narrow ditch that follows the outer edge of all of the inner FR4 layers. This is the o-ring housing to render the active part of the detector gas tight. The outer faces of the top and the bottom inner openings should be covered with a thin film to keep the Ar-CO2 gas mixture in and the water vapor from the surroundings out while letting in as many of the incoming particles. The topmost layer's inner face will house the first layer of cathode wires. Following the literature, Copper wires with 2% beryllium content will be installed with 2mm pitch. The upper side of the next layer houses the ditches to match the soldering joints of the first cathode wires, and its lower side will house the anode wires. These are selected to be Gold plated tungsten, with 4mm separation. The third layer is a copy of the uppermost one, except it has ditches on the upper side to match the soldering joints of the anode wires. The last (bottommost) layer contains the usual soldering joint protection ditches on its upper side. The ditches are 1mm deep to host PCB segments with appropriate hole spacing, which is 2mm for both anode and cathode planes. The connector side is made to extend out of the FR4 frame. It is to host a specially designed PCB that will convert the 2mm pitch on the wire side to 2.54mm (0.1 inch) on the connector side. The design of this connector, made using EAGLE (free for educational use) can be seen in below Figure.

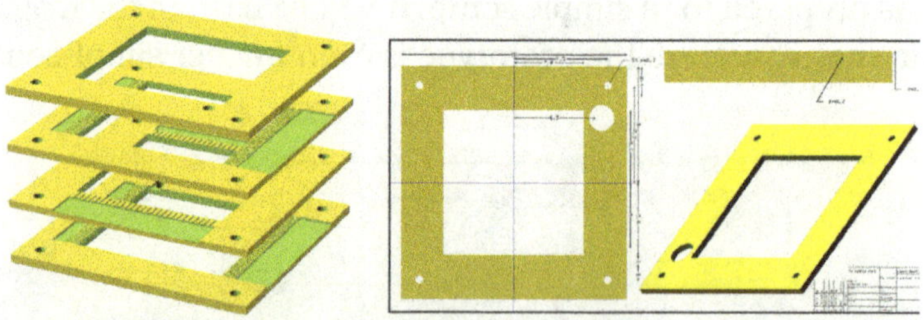

Figure: Delay Wire Chamber 3D design on the left and engineering drawing on the right.

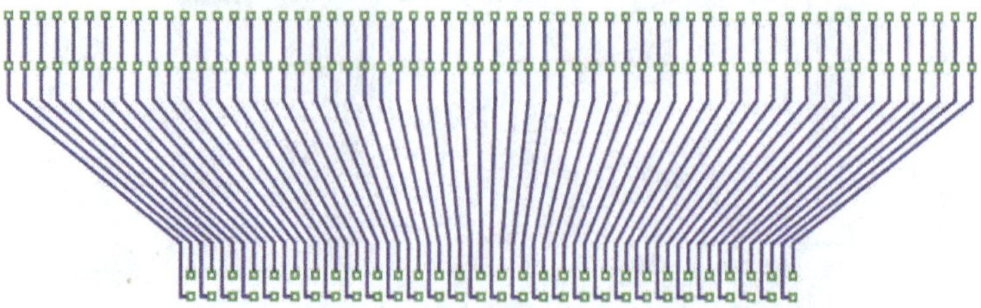

Figure: Pitch converter from 2mm to 0.1 inch.

The electronic readout circuit design with appropriate delays between the anode wires has been around for more than 20 years. Unfortunately some of the integrated circuits are not available anymore. A new circuit design with smaller footprint and with more up-to-date circuits is in progress, and will be used.

Machining and Construction

The FR4 frames were initially processed with a simple drilling machine with mechanical control wheels (which can be available even in high schools), all four stacked together as in Fig. 4 left side. After rendering of the common design features such as the inner window and the main screw holes, each layer is machined individually based on the design. The resulting four planes can be seen on the same figure right side. The same photo contains the gas connectors and the o-rings for the main screws. Note that due to the simplicity of the tools, the ditches have rounded corners to which the PCBs will have to be adapted.

After completion of the frames, the next task is the attachment of various components such as the gas input output connectors and wire holding PCBs. These pieces are glued to the FR4 material using epoxy paste which would not contaminate the ionization gas. Epoxy paste is strong enough to keep the wires in tension when the PCBs are glued and it can be found in local hardware stores or in big markets. Instead of using a machine to ensure the same mechanical tension across the conductive wires, a simple method is used: one side of the wire is connected to the PCB and the other side is connected to a fishing weight of about 100gr. These weights can be found in local sports store and can be pre-adjusted to the same mass with a simple kitchen scale, which is usually accurate to a gram. Fig. 5 contains the second cathode plane with two wires installed on each side. Note that the wire pitch converter is on the left hand side. At the outer part of the pitch converter, a standard 60 pin connector is soldered.

Figure: Machining the FR4 with a drilling machine with simple controls

Figure: Installing the wires

After installing the wires on the PCBs, the thin film should be glued to the upper and lower FR4 layers. For this purpose Kapton film with 25 micron thickness is selected, as it is durable under pressure and stable under temperature changes . Instead of Kapton, one can also use Mylar which may be easier to procure. However, Kapton is preferred for its even thinner sheet availability (down to 13 microns), which would allow even less energetic particles to enter the chamber. Fluka simulations show that protons of 1.3 MeV can travel about 30 microns inside the Kapton film, while for 1.4 MeV protons the path length increases to 33 microns [10, 11]. For gas tightness, o-rings of 5mm width and about 1mm thickness can be used. To drive down the cost of the detector, it is envisaged to produce this square o-ring from the packaging rubbers of appropriate shape and size. In this project, 20cm long big rubber bands are cut to be fit in the o-ring ditches and glued at each end with a strong glue, forming a complete square of side 16cm. After fitting the o-rings, all the layers are stacked consecutively allowing the cathode wires to be parallel and the anode wires perpendicular in between. Then screws of 6mm are sticked at four corners and tightened.

Detector Tests

First item to be tested is gas tightness since a certain level of gas pureness needs to be achieved. One also needs to be careful about the materials as some of the materials like the glue may easily cause contamination in the gas. If the gas in the detector is not pure enough, then when the high voltage is applied, the wires might produce sparks which in turn, may cause harm to the detector or the electronic circuitry. To test gas tightness, when all the individual construction work is finished but before installing the wires, the assembled chamber is filled with gas. By applying a leak detector spray, possible leaks are examined. Leak detection can also be done with simple soap foam but then it should be cleaned very carefully. In case of any leak, one can change the o-rings or apply more epoxy paste locally. Only after being sure that the detector is gas tight, and the installation of the wires is complete, the high voltage can be applied by increasing it slowly and observing the detector for any sparks. After the gas tightness and the high voltage tests, the next test is for electronic circuitry. By sending analog signals from a signal generator to a channel in the readout system, it is possible to observe the final signal from the oscilloscope. The final two signals should be compatible with the expected values depending on the channel and the delay step of the microchips used. At this point the detector becomes ready for particles, in particular cosmic particles (muons) for the first tests. The cosmic muons should be equally distributed on the DWC active area. Further studies could be made with the new detector installed between two previously calibrated and tested DWCs. Again testing with cosmic muons, one should be able to find the track of the cosmic particles using the older DWCs and check the efficiency of the new detector by comparing its hit position to the fitted track.

Scintillation Counters

The scintillation counter is a device which measures the ionizing radiation. It consists of sensor, which is also known as Scintillator, which fluoresces when it comes in contact of ionizing radiations. This light is then sensed by a photomultiplier tube (PMT). This PMT is attached to sophisticated electronic circuitry which counts and measures the radiation by amplifying the signals received from the PMT.

In 1944, when Sir Samuel Curran was experimenting on the works of Antoine Henri Becquerel to improve it, he invented the scintillation counter. Earlier to the invention of Scintillation counter, the Spinthariscope was used to detect the scintillation event by placing it on eye.

Operation of Scintillation Counter

The below shown picture contains main components of the Scintillation Counter.

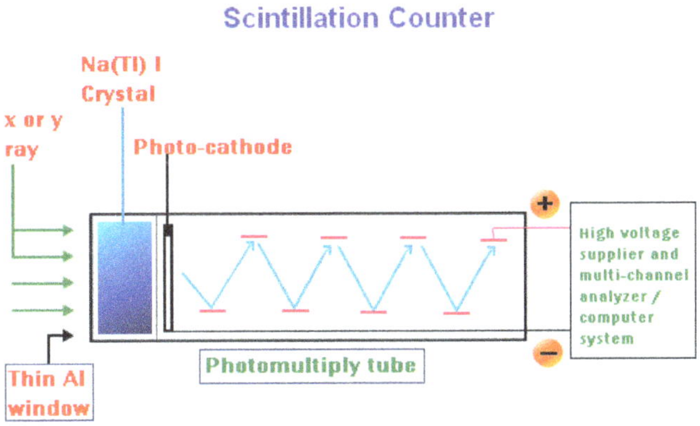

The one end of the scintillation counter has scintillation phosphor, like thalium activated sodium iodide crystal. This crystal is used as an sensor which senses the radiation. This sensed radiation is then detected by the PMT (Photomultiply tube), which produces the proportional pulses of electric current. This current is then measured by the multi channel analyzer.

Scintillation Phosphors are Majorly of three Types

- Inorganic Crystals : The inorganic crystals have high atomic number and density. Gamma rays are detected by these types of the crystal very efficiently. In these crystals the pulses get decayed in approx 1 micro seconds. The common examples of inorganic crystals are sodium iodide, zinc sulphide, lithium iodide, etc.
- Organic Crystals : These types of crystals are generally employed in the scintillation counter which is used for the detection of the beta rays. They exhibit the pulse decay time of 10 nano seconds. The common examples of organic crystals are stilbene, anthracene, etc.
- Plastic Phosphors : These types of crystals are used in the fast neutron detectors. They have very high hydrogen content and hence are the ideal fast neutron detectors. The pulse decay in these types of crystal is 1 or 2 nano seconds.

Liquid Scintillation Counter

The Liquid Scintillation Counters are majorly used in Life Sciences for measuring the betaemissions from the nuclides. The liquid scintillation counter uses a mixture of solvent and fluors, the mixture is often referred to as cocktail. The sample is added to the cocktail. The sample when emits a beta particle, the energy is transferred to the solvent. The solvent, in turn, transfer this energy to the fluors. The fluors receiving the energy get excited to the unstable state. It releases the extra energy and come down to the stable state. This energy is detected and beta radiations are measured.

Few additives are added to these cocktails so as to shift the wavelength of emitted light in a region where it can be easy to detect.

The liquid scintillation counter has two PMTs for better and efficient beta particles detection. There are certain factors which interferes with the beta particles detections, these are:

- Optical Cross Talk: This occurs when certain events in one tube induces pulse in the other PMT tube. This factor is overcome by using sophisticated circuitry which can differentiate the events occurred due to sample or due to non sample.
- Line Transmission Noise: A faulty line may create a false detection and hence the experiment fails. To overcome such factor the electric line should be free from the noises by using good filters and equalisers.
- Static Induced Noise: The static charges are produced all around the PMT due to the friction of the vials containing the sample. For overcoming this factor the vials must be properly grounded.
- Natural Radioactivity or Cosmic Radiation: The natural radioactivity is the main cause for the spurious count of the particles. The natural radioactive emission interferes with the count by triggering the false electric pulses which could be counted. To overcome this factor the experimental setup should be properly shielded from the active and/or passive shielding materials.
- Radio Frequency Interference: These are common interferes. The source of RF could be any motor or light or switch. To overcome the interference caused by them is to use a RF circuitry which detects these types of Interference and removed them from triggering false pulse count in the counter.
- Apart from these factors there are several other factors which can affect the counting of the scintillation counter. In short, the liquid scintillation device is very sensitive and hence utmost care should be taken to avoid the false triggering of the pulses in the counter.

Scintillation Counting

The Scintillation Counting is based on count the pulses produced by the fluorescence of the scintillation material. The scintillation counter is composed of scintillator and the photo detector devices. The scintillators fluoresce when they come in contact of the high energy ionizing particle like beta or gamma rays. The electric pulses are produced by the scintillator which are proportional to the amount of the fluoresce.

Applications of the Scintillation Counter

The scintillation counters are used in several places to detect the radiations. Some of the examples are,

- As a Portable survey meters.
- Used in medical imaging.
- Used as border security devices to detect any radiation from the enemy side.

- Used as a monitoring device in the nuclear plant to monitor and trace the leakage of the nuclear radiation from the plant.
- Used to detect the radon in the various day to day used materials especially water. The radon is radioactive, odourless, colourless noble gas and can get mixed easily with any substance.
- In the industry to detect alpha or beta or gamma radiations. The industrial counter are made up of different types of fluoresce. One counter may contain two different fluoresce, like one for alpha particle detection and another for beta particle detection.
- They are used as a spectrometers also, since they contains PMT which converts a high energy photon into several photon with lesser energy. These pulses could then be sorted with respect to their height and the total energy per height is calculated to approximate the energy spectrum of the incident radaitons.

Semiconductor Solid State Detectors

Solid-state detector or semiconductor radiation detector is a radiation detector in which a semiconductor material such as a silicon or germanium crystal constitutes the detecting medium. One such device consists of a *p-n* junction across which a pulse of current develops when a particle of ionizing radiation traverses it. In a different device, the absorption of ionizing radiation generates pairs of charge carriers (electrons and electron-deficient sites called holes) in a block of semiconducting material; the migration of these carriers under the influence of a voltage maintained between the opposite faces of the block constitutes a pulse of current. The pulses created in this way are amplified, recorded, and analyzed to determine the energy, number, or identity of the incident-charged particles. The sensitivity of these detectors is increased by operating them at low temperatures—commonly that of liquid nitrogen, −164 °C (−263 °F)—which suppresses the random formation of charge carriers by thermal vibration.

Semiconductor Detector Operational Characteristics

Effect of bias voltage on detector performance: In order to produce an electric field large enough to achieve a charge carrier efficient collection, a reverse bias voltage of hundreds or thousands of volts has to be applied between the semiconductor detector electrodes. In fact, if low bias voltage is imposed, the pulse height deriving from radiations that are fully stopped within the depletion region rises with the applied voltage. This variation is due to an incomplete carrier charge collection as consequence of trapping/recombination mechanisms along the incident particle track. The fraction escaping collection, decreases as the electric field is increased.

When a reverse voltage is applied to a junction detector, a current of the order of a fraction of microampere or nanoampere is usually observed (depending on the particular semiconductor material adopted) even in absence of ionizing radiation. The origin of this leakage current are related both to the bulk volume and surface of the detector.

Sources of the bulk leakage current are both the minority carriers, that are conducted across the junction because of the reverse voltage applied, and the thermal generation of electron/hole pairs within the depletion region. The first contribute is generally small while the second depends on the operation temperature. For example, silicon and gallium arsenide detectors have a sufficient low thermally generated leakage current to allow their use at room temperature while germanium devices must operate at reduced temperature as consequence of the lower energy gap.

Surface leakage current takes place at the junction edges where voltage gradients are present. It depends on various factors such as humidity and detector surface contamination.

Random fluctuations that occur in leakage current could tend to obscure the small signal current that flows following an ionizing event and could represent a noise source in many situations. Methods for leakage current reduction, such as guard-rings, have to be taken into account in semiconductor detector design phase.

Pulse Rise Time

Semiconductor detector timing properties can be evaluated taking into account the output signal rise time. It is composed of the charge transit time and the plasma time contributes.

The charge transit time is related to the motion of electron/hole pairs created by a radiation beam hitting the detector depletion region. Therefore, it depends on the electric field intensity and the spatial charge zone width. In fully depleted detectors, the semiconductor physical thickness fixes the depletion width and consequently, the electric field growth reduces the rise time. In partially depleted detectors, the spatial charge region increases with increasing the applied voltage and therefore the effect of a high voltage bias is to increase both the electric field and the distance over which charge pairs must be collected.

The relation of the charge transit time on the applied bias voltage is enough complicated even if it can be simplified if suitable assumptions are made.

When, the incident radiation is composed of heavy charged particles, the plasma time component is observed. In fact, in this situation the electron/hole pair density along the track of the hitting particles is sufficient high to form a plasma (like charge cloud) that shields the interior from the electric field influence. Only the charge carriers at the outer edge of the cloud migrate because they are under the influence of the electric field. The outer regions are gradually eroded away till the inner charges are subjected to the applied electric field and they drift. Therefore, the plasma time can be defined as the time required for the charge cloud to disperse to the point where normal charge collection proceeds.

Dead layer and channeling: When semiconductor detectors are irradiated with weakly penetrating particles, such as heavy charged particles, an high amount of energy can be lost before the ray reaches the detector active volume. The thickness of this dead layer is, generally, a function of the applied voltage.

When the hitting ray is perpendicular to the detector surface (i.e., the incidence angle is zero), the energy loss in the dead layer is:

$$\Delta E(0) = \frac{dE(0)}{dx} d$$

where, d is the dead layer thickness.

The energy loss for a generic β incidence angle is:

$$\Delta E(\beta) = \frac{\Delta E(\beta)}{\cos\beta}$$

Moreover, in crystalline materials the energy loss rate of charged particles depends on their path orientation; in fact particles travelling parallel to detector crystal planes are characterized by an energy loss rate lower that that of particles directed in arbitrary directions. These particles (sometimes called channelled particles) penetrate significantly farther through the crystal.

Radiation Damage

The radiation process produces inside the detector some damages whose nature depends on the particular particle hitting the semiconductor material. The radiation induced damages can be classified in two categories: bulk and surface type.

The impinging radiation can interact with semiconductor material nuclei producing an irreversible damage. In this situation, a bulk damage is produced (called Frenkel defect) because a material atom is displaced from its normal lattice site. The generated vacancy and the original atom, now at an interstitial position, act as:

- Recombination/generation centres because they are able both to capture and to emit electrons and holes. This alternate emission of electrons and holes inside the detector depleted region increases the leakage current
- Trapping centres because electrons and holes are captured and re-emitted. The trapped signal charge could be released too late producing an incomplete charge collection at the detector electrodes

Moreover, they could change the charge density in the depleted zone and consequently, an increased bias voltage is required to make the detector fully sensitive.

Therefore, when enough of these defects are generated, carrier lifetime and charge collection efficiency are reduced while detector energy resolution is degraded because of fluctuations in the charge lost amount.

After the end of the irradiation process, it is possible to notice that the produced damage to the detector reduces with time. This effect is called annealing. The rate of damage decrease is strongly dependent on the temperature at which the detector is kept during the waiting period. True annealing, in which the crystal becomes perfect again, does exist but in many cases defects may just be transformed into other more stable defect types with changed properties.

Surface damages involve charge build up either at interface or in overlaying electrically insulating layers (i.e., a thermal oxide on silicon) and can cause heavy surface inversion or accumulation. In fact, energy absorbed by electronic ionization in insulating layers liberates charge carriers which diffuse or drift to other locations where they are trapped, leading to unintended concentrations of charge and as a consequence, parasitic fields.

Therefore, oxide damage is caused by ionizing irradiation such as photons, x-rays and charged particles, while semiconductor bulk damage, requiring damage of the crystal structure, is generated by massive particles such as neutrons and protons .

Semiconductor Detector Classification

Existing semiconductor detectors differ from one another because of the adopted material or the method by which the material is treated.

In the following sections, construction methods and operating characteristics of the most popular semiconductor detectors are briefly described.

Surface Barrier Detectors

A p-n junction is created on a n type semiconductor by evaporation of a thin gold layer which acts as an acceptor and simultaneous serves as an electrode. Best results are obtained if the evaporation phase is carried out under conditions that produce surface oxidation. In fact, as a result of surface oxidization, surface energy states are produced that induce a hole high density so to form a p type layer on the surface. Moreover, the oxidation layer reduces the surface leakage current improving the detector performance.

Hence, the front electrode operates as a rectifying contact, while the back electrode serves only as an ohmic contact.

Diffused and Implanted Junction Detectors

A homogeneous crystal of p-type material is the basic material for this detector types.

One surface of the wafer is doped by diffusing n type impurities at high temperature (typically phosphorus) so to create a thin n type semiconductor layer. Therefore, a p-n junction is formed which has a depleted zone extended primarily into the semiconductor p type zone as the n type surface layer is heavily doped compared with the p type crystal. For the reasons previously described, much of the surface layer is outside the depletion region and it represents a dead layer. In some applications, the dead layer presence is a disadvantage as an amount of particle energy is lost before the detector active zone is reached.

An alternative method to introduce doping impurities at semiconductor surface is the ion implantation method which makes the concentration profile of the added impurity controllable by changing the incident ion energy. To reduce the radiation damage caused by the incident ions, a subsequent annealing step is necessary which is characterized by a process temperature lower than the one which is needed for the diffusion of dopants to form a diffused junction. For this reason, a reduced defect number is produced and consequently, a lower leakage current is expected in comparison with diffused barriers.

Lithium Drifted Junction Detectors

Using conventional junction techniques it is not possible obtaining very wide depletion layers. To provide detectors of greater sensitive volume, a process has been devised which creates a region of compensated or intrinsic semiconductor in which donor and acceptor concentration are balanced. These detectors are called lithium drifted silicon detector or lithium drifted germanium detector and designed as Si(Li) or Ge(Li).

During the fabrication process, lithium is diffused through one surface of a p type crystal where it acts as a donor impurity. The resulting p-n junction is reverse biased while the crystal temperature is increased to enhance the mobility of lithium donors. Lithium donors are slowly drawn by the electric field into the p type semiconductor where their concentration will increase and approach that of the original acceptor impurities. Thus, a region is created that looks like an intrinsic semiconductor.

Once the drifting process is completed, the detector has the configuration indicated in Fig. 13.

The process of compensating impurities by drifting lithium into a semiconductor, has been developed to produce either silicon and germanium devices.

More recently, detectors having wide active volume are possible adopting the high purity germanium configuration (HPGe) and consequently, Ge(Li) is no longer popular. However the technique remains important in silicon.

Compound Semiconductor Detectors

Besides semiconductors composed of a single chemical element (such as Si and Ge), various semiconductors exist in which atoms of a different element are able to substitute a particular material without altering its crystal structure.

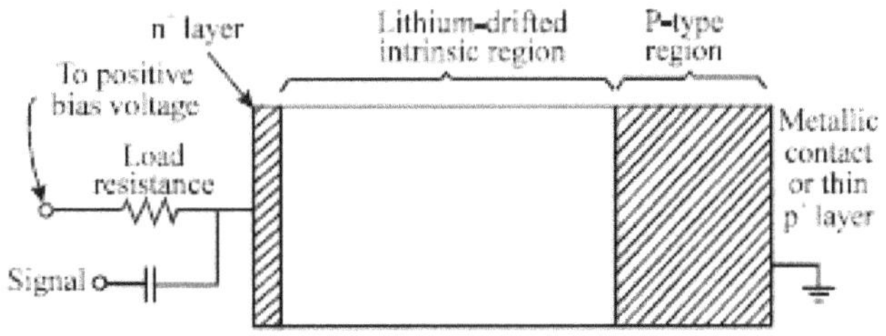

Fig: Lithium drifted junction detector configuration

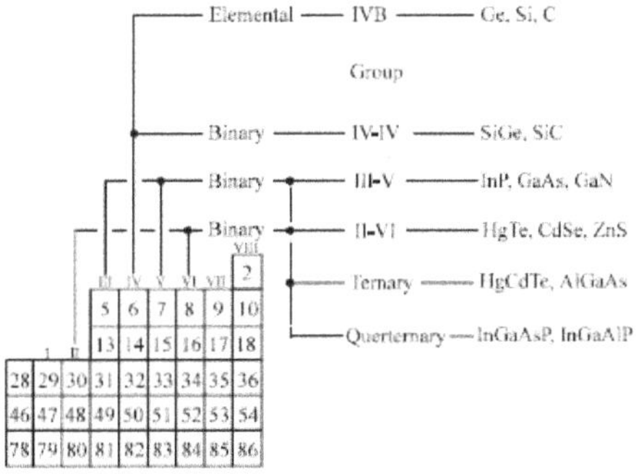

Figure: Compound semiconductor list

A list of this materials, called compound semiconductors, is shown in figure where, the main group and their composing elements are indicated.

In particular, compound semiconductors have many advantages arising from wide range of stopping powers to band gaps available. Mixing and matching available band gaps and stopping powers appropriately, different radiation detectors for specific applications can be realized. For example, an energy gap closed to 1.5eV, minimizes leakage current at room temperature and provides an adequate electron-hole pairs per absorbed photon. Moreover, the choice of materials with very high stopping powers makes the realization of thin detectors possible with resulting benefits in radiation tolerance and leakage currents. Unlike, materials characterized by very small stopping powers increase the detector scattering efficiency, which is a basic requirement for polarimetry measurements. A suitable choice of semiconductors having appropriate band gap and stopping power values, allows the fabrication of detectors with a very wide dynamic range.

However, compound semiconductors have some limitations such as poor mobility or reduced carrier lifetime of one or both charge carriers. Therefore, the mobility-lifetime product, which represents a classical detector figure of merit, is for these materials lower than the corresponding value of the elemental semiconductors. The reason can be attribute to trapping centers caused by impurities, lack of stoichiometry, plastic deformations due to mechanical damage during fabrication processes.

References

- Bubble-chamber: encyclopedia2.thefreedictionary.com, Retrieved 10 March 2018
- How-Gas-Detectors-Work, instruments-controls: thomasnet.com, Retrieved 16 April 2018
- Das Gupta, N. N.; Ghosh S. K. (1946). "A Report on the Wilson Cloud Chamber and its Applications in Physics". Reviews of Modern Physics. 18 (2): 225–365. Bibcode:1946RvMP...18..225G. doi:10.1103/RevModPhys.18.225
- Scintillation-counter, modern-physics: tutorvista.com, Retrieved 23 June 2018
- Solid-state-detector, science: britannica.com, Retrieved 09 July 2018

6

Nuclear Fusion and Fission

Nuclear fusion is the reaction characterized by the fusion of two or more atomic nuclei. In contrast, nuclear fission refers to the nuclear decay process in which an atomic nucleus splits into smaller elements, along with the release of a massive amount of energy. The topics discussed in this chapter include thermonuclear fusion, controlled fusion, fusion reaction, Maxwell-averaged fusion reactivities, etc. which will provide a thorough understanding of nuclear reactions.

Nuclear Fusion

In physics and nuclear chemistry, nuclear fusion is the process by which multiple atomic particles join together to form a heavier nucleus. It is accompanied by the release or absorption of energy. Iron and nickel nuclei have the largest binding energies per nucleon of all nuclei and therefore are the most stable. The fusion of two nuclei lighter than iron or nickel generally releases energy, while the fusion of nuclei heavier than iron or nickel absorbs energy. The opposite is true for nuclear fission. Nuclear fusion is naturally found in stars.

Fusion reactions power the stars and produce all but the lightest elements in a process called nucleosynthesis. Whereas the fusion of light elements in the stars releases energy, production of the heaviest elements absorbs energy, so it can only take place in the extremely high-energy conditions of supernova explosions.

When the fusion reaction is a sustained uncontrolled chain, it can result in a thermonuclear explosion, such as what is generated by a hydrogen bomb. Reactions that are not self-sustaining can still release considerable energy, as well as large numbers of neutrons.

Research into controlled fusion, with the aim of producing fusion power for the production of electricity, has been conducted for over 50 years. It has been accompanied by extreme scientific and technological difficulties, and as of yet has not been successful in producing workable designs. As of the present, the only self-sustaining fusion reactions produced by humans have been produced in hydrogen bombs, where the extreme power of a fission bomb is necessary to begin the process. While some plans have been put forth to attempt to use the explosions of hydrogen bombs to generate electricity (e.g. PACER), none of these have ever moved far past the design stage.

It takes considerable energy to force nuclei to fuse, even those of the lightest element, hydrogen. This is because all nuclei have a positive charge (due to their protons), and as like charges repel, nuclei strongly resist being put too close together. Accelerated to high speeds (that is, heated to thermonuclear temperatures), however, they can overcome this electromagnetic repulsion and get close enough for the strong nuclear force to be active, achieving fusion. The fusion of lighter nuclei, creating a heavier nucleus and a free neutron, will generally release more energy than it took to force them together—an exothermic process that can produce self-sustaining reactions.

The energy released in most nuclear reactions is much larger than that in chemical reactions, because the binding energy that holds a nucleus together is far greater than the energy that holds electronsto a nucleus. For example, the ionization energy gained by adding an electron to a hydrogen nucleus is 13.6 electron volts—less than one-millionth of the 17 MeV released in the D-T (deuterium-tritium) reaction shown to the top right. Fusion reactions have an energy density many times greater than nuclear fission—that is, per unit of mass the reactions produce far greater energies, even though *individual* fission reactions are generally much more energetic than *individual*-fusion reactions—which are themselves millions of times more energetic than chemical reactions. Only the direct conversion of mass into energy, such as with collision of matter and antimatter, is more energetic per unit of mass than nuclear fusion.

Building upon the nuclear transmutation experiments of Ernest Rutherford done a few years earlier, fusion of light nuclei (hydrogen isotopes) was first observed by Mark Oliphant in 1932, and the steps of the main cycle of nuclear fusion in stars were subsequently worked out by Hans Bethe throughout the remainder of that decade. Research into fusion for military purposes began in the early 1940s, as part of the Manhattan Project, but was not successful until 1952. Research into controlled fusion for civilian purposes began in the 1950s, and continues to this day.

Requirements

A substantial energy barrier must be overcome before fusion can occur. At large distances two naked nuclei repel one another because of the repulsive electrostatic force between their positively charged protons. If two nuclei can be brought close enough together, however, the electrostatic repulsion can be overcome by the nuclear force which is stronger at close distances.

When a nucleon such as a proton or neutron is added to a nucleus, the nuclear force attracts it to other nucleons, but primarily to its immediate neighbors due to the short range of the force. The nucleons in the interior of a nucleus have more neighboring nucleons than those on the surface. Since smaller nuclei have a larger surface area-to-volume ratio, the binding energy per nucleon due to the strong force generally increases with the size of the nucleus but approaches a limiting value corresponding to that of a fully surrounded nucleon.

The electrostatic force, on the other hand, is an inverse-square force, so a proton added to a nucleus will feel an electrostatic repulsion from *all* the other protons in the nucleus. The electrostatic energy per nucleon due to the electrostatic force thus increases without limit as nuclei get larger.

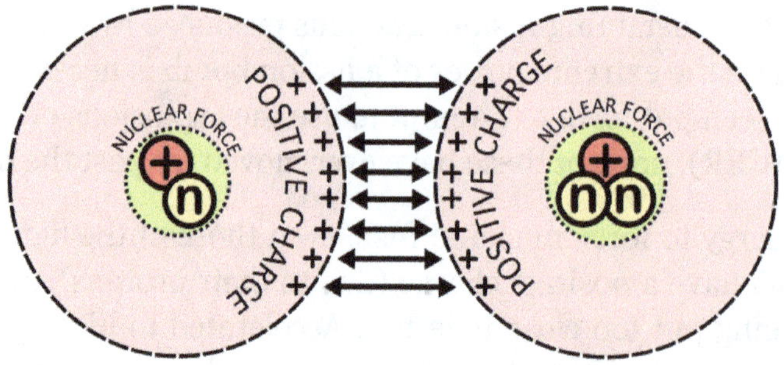

The electrostatic force caused by positively charged nuclei are very strong over long distances, but at short distances the nuclear force is stronger. As such, the main technical difficulty for fusion is getting the nuclei close enough to fuse (distances not to scale).

The net result of these opposing forces is that the binding energy per nucleon generally increases with increasing size, up to the elements iron and nickel, and then decreases for heavier nuclei. Eventually, the binding energy becomes negative and very heavy nuclei are not stable. The four most tightly bound nuclei, in decreasing order of binding energy, are $^{62}Ni, ^{58}Fe, ^{56}Fe,$ and ^{60}Ni. Even though the nickel isotope ^{62}Ni is more stable, the iron isotope ^{56}Fe is an order of magnitude more common. This is due to a greater disintegration rate for ^{62}Ni in the interior of stars driven by photon absorption.

A notable exception to this general trend is the helium-4 nucleus, whose binding energy is higher than that of lithium, the next heavier element. The Pauli exclusion principle provides an explanation for this exceptional behavior—it says that because protons and neutrons are fermions, they cannot exist in exactly the same state. Each proton or neutron energy state in a nucleus can accommodate both a spin up particle and a spin down particle. Helium-4 has an anomalously large binding energy because its nucleus consists of two protons and two neutrons; so all four of its nucleons can be in the ground state. Any additional nucleons would have to go into higher energy states.

The situation is similar if two nuclei are brought together. As they approach each other, all the protons in one nucleus repel all the protons in the other. Not until the two nuclei actually come in contact can the strong nuclear force take over. Consequently, even when the final energy state is lower, there is a large energy barrier that must first be overcome. It is called the Coulomb barrier.

The Coulomb barrier is smallest for isotopes of hydrogen—they contain only a single positive charge in the nucleus. A bi-proton is not stable, so neutrons must also be involved, ideally in such a way that a helium nucleus, with its extremely tight binding, is one of the products.

Using deuterium-tritium fuel, the resulting energy barrier is about 0.01 MeV. In comparison, the energy needed to remove an electronfrom hydrogen is 13.6 eV, about 750 times less energy. The (intermediate) result of the fusion is an unstable ^{5}He nucleus, which immediately ejects a neutron with 14.1 MeV. The recoil energy of the remaining ^{4}He nucleus is 3.5 MeV, so the total energy liberated is 17.6 MeV. This is many times more than what was needed to overcome the energy barrier.

If the energy to initiate the reaction comes from accelerating one of the nuclei, the process is called *beam-target* fusion; if both nuclei are accelerated, it is *beam-beam* fusion. If the nuclei are part of a plasma near thermal equilibrium, one speaks of *thermonuclear* fusion. Temperature is a measure of the average kinetic energy of particles, so by heating the nuclei they will gain energy and eventually have enough to overcome this 0.01 MeV. Converting the units between electron-volts and Kelvin shows that the barrier would be overcome at a temperature in excess of 120 million Kelvin—a very high temperature.

There are two effects that lower the actual temperature needed. One is the fact that temperature is the *average* kinetic energy, implying that some nuclei at this temperature would actually have much higher energy than 0.01 MeV, while others would be much lower. It is the nuclei in the high-energy tail of the velocity distribution that account for most of the fusion reactions. The other effect is quantum tunneling. The nuclei do not actually have to have enough energy to overcome the Coulomb barrier completely. If they have nearly enough energy, they can tunnel through the remaining barrier. For this reason fuel at lower temperatures will still undergo fusion events at a lower rate.

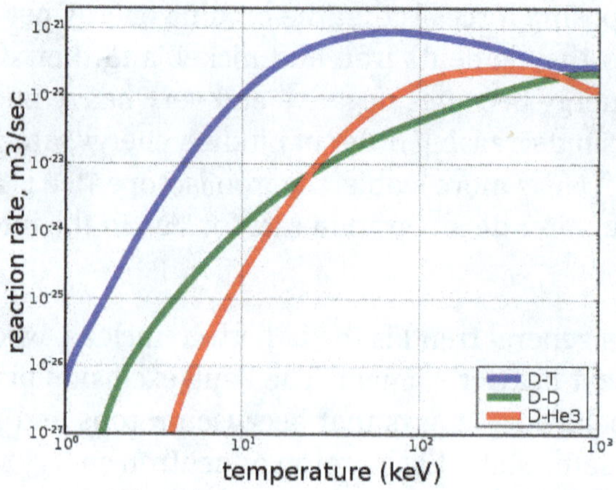

The fusion reaction rate increases rapidly with temperature until it maximizes and then gradually drops off. The DT rate peaks at a lower temperature (about 70 keV, or 800 million Kelvin) and at a higher value than other reactions commonly considered for fusion energy.

The reaction cross section σ is a measure of the probability of a fusion reaction as a function of the relative velocity of the two reactant nuclei. If the reactants have a distribution of velocities, e.g. a thermal distribution with thermonuclear fusion, then it is useful to perform an average over the distributions of the product of cross section and velocity. The reaction rate (fusions per volume per time) is <σv> times the product of the reactant number densities:

$$f = n_1 n_2 \langle \sigma v \rangle$$

If a species of nuclei is reacting with itself, such as the DD reaction, then the product $n_1 n_2$ must be replaced by $(1/2)n^2$.

$\langle \sigma v \rangle$ increases from virtually zero at room temperatures up to meaningful magnitudes at temperatures of 10–100 keV. At these temperatures, well above typical ionization energies (13.6 eV in the hydrogen case), the fusion reactants exist in a plasma state.

The significance of <σv> as a function of temperature in a device with a particular energy confinement time is found by considering the Lawson criterion.

Fuel Confinement Methods

Gravitational

One force capable of confining the fuel well enough to satisfy the Lawson criterion is gravity. The mass needed, however, is so great that gravitational confinement is only found in stars (the smallest of which are brown dwarfs). Even if the more reactive fuel deuterium were used, a mass greater than that of the planet Jupiter would be needed.

Magnetic

Since plasmas are very good electrical conductors, magnetic fields can also confine fusion fuel. A variety of magnetic configurations can be used, the most basic distinction being between mirror confinement and toroidal confinement, especially tokamaks and stellarators.

Inertial

A third confinement principle is to apply a rapid pulse of energy to a large part of the surface of a pellet of fusion fuel, causing it to simultaneously "implode" and heat to very high pressure and temperature. If the fuel is dense enough and hot enough, the fusion reaction rate will be high enough to burn a significant fraction of the fuel before it has dissipated. To achieve these extreme conditions, the initially cold fuel must be explosively compressed. Inertial confinement is used in the hydrogen bomb, where the driver is x-rayscreated by a fission bomb. Inertial confinement is also attempted in "controlled" nuclear fusion, where the driver is a laser, ion, or electron beam, or a Z-pinch.

Some other confinement principles have been investigated, such as muon-catalyzed fusion, the Farnsworth-Hirsch fusor and Polywell (inertial electrostatic confinement), and bubble fusion.

Production Methods

A variety of methods are known to effect nuclear fusion. Some are "cold" in the strict sense that no part of the material is hot (except for the reaction products), some are "cold" in the limited sense that the bulk of the material is at a relatively low temperature and pressure but the reactants are not, and some are "hot" fusion methods that create macroscopic regions of very high temperature and pressure.

Locally Cold Fusion

Muon-catalyzed fusion is a well-established and reproducible fusion process that occurs at ordinary temperatures. It was studied in detail by Steven Jones in the early 1980s. It has not been reported to produce net energy. Net energy production from this reaction is not believed to be possible because of the energy required to create muons, their 2.2 μs half-life, and the chance that a muon will bind to the new alpha particle and thus stop catalyzing fusion.

Generally Cold, Locally Hot Fusion

Accelerator based light-ion fusion. Using particle accelerators it is possible to achieve particle kinetic energies sufficient to induce many light ion fusion reactions. Of particular relevance into this discussion are devices referred to as sealed-tube neutron generators. These small devices are miniature particle accelerators filled with deuterium and tritium gas in an arrangement which allows ions of these nuclei to be accelerated against hydride targets, also containing deuterium and tritium, where fusion takes place. Hundreds of neutron generators are produced annually for use in the petroleum industry where they are used in measurement equipment for locating and mapping oil reserves. Despite periodic reports in the popular press by scientists claiming to have invented "table-top" fusion machines, neutron generators have been around for half a century. The sizes of these devices vary but the smallest instruments are often packaged in sizes smaller than a loaf of bread. These devices do not produce a net power output.

In sonoluminescence, acoustic shock waves create temporary bubbles that collapse shortly after creation, producing very high temperatures and pressures. In 2002, Rusi P. Taleyarkhan reported the possibility that bubble fusion occurs in those collapsing bubbles (sonofusion). As of 2005, experiments to determine whether fusion is occurring give conflicting results. If fusion is occurring, it is because the local temperature and pressure are sufficiently high to produce hot fusion.

The Farnsworth-Hirsch Fusor is a tabletop device in which fusion occurs. This fusion comes from high effective temperatures produced by electrostatic acceleration of ions. The device can be built inexpensively, but it too is unable to produce a net power output.

Antimatter-initialized fusion uses small amounts of antimatter to trigger a tiny fusion explosion. This has been studied primarily in the context of making nuclear pulse propulsion feasible. This is not near becoming a practical power source, due to the cost of manufacturing antimatter alone.

Pyroelectric fusion was reported in April 2005 by a team at UCLA. The scientists used a pyroelectric crystal heated from −34 to 7 °C (−30 to 45 °F), combined with a tungsten needle to produce an electric field of about 25 gigavolts per meter to ionize and accelerate deuterium nuclei into an erbium deuteride target. Though the energy of the deuterium ions generated by the crystal has not been directly measured, the authors used 100 keV (a temperature of about 10^9 K) as an estimate in their modeling. At these energy levels, two deuterium nuclei can fuse together to produce a helium-3 nucleus, a 2.45 MeV neutron and bremsstrahlung. Although it makes a useful neutron generator, the apparatus is not intended for power generation since it requires far more energy than it produces.

Hot Fusion

"Standard" "hot" fusion, in which the fuel reaches tremendous temperature and pressure inside a fusion reactor or nuclear weapon.

The methods in the second group are examples of non-equilibrium systems, in which very high temperatures and pressures are produced in a relatively small region adjacent to material of much lower temperature. In his doctoral thesis for MIT, Todd Rider did a theoretical study of all quasineutral, isotropic, non-equilibrium fusion systems. He demonstrated that all such systems will leak energy at a rapid rate due to bremsstrahlung, radiation produced when electrons in the plasma hit other electrons or ions at a cooler temperature and suddenly decelerate. The problem is not as pronounced in a hot plasma because the range of temperatures, and thus the magnitude of the deceleration, is much lower. Note that Rider's work does not apply to non-neutral and/or anisotropic non-equilibrium plasmas.

Important Reactions

Astrophysical Reaction Chains

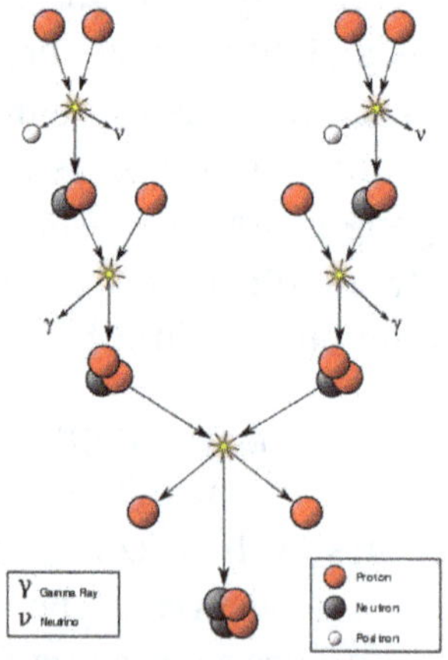

The proton-proton chain dominates in stars the size of the sun or smaller

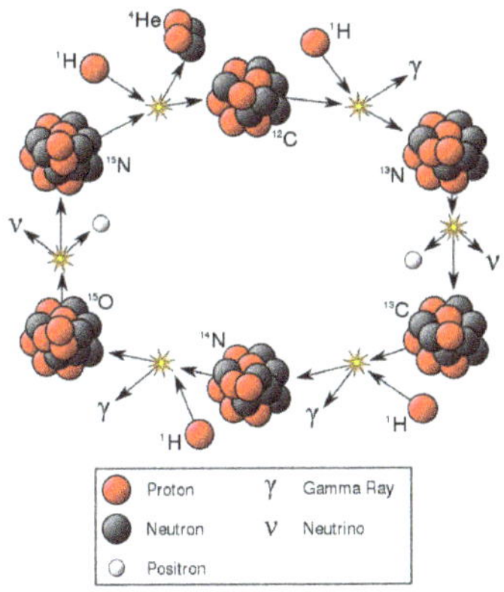

The CNO cycle dominates in stars heavier than the sun

The most important fusion process in nature is that which powers the stars. The net result is the fusion of four protons into one alpha particle, with the release of two positrons, two neutrinos (which changes two of the protons into neutrons), and energy, but several individual reactions are involved, depending on the mass of the star. For stars the size of the sun or smaller, the proton-proton chain dominates. In heavier stars, the CNO cycle is more important. Both types of processes are responsible for the creation of new elements as part of stellar nucleosynthesis.

At the temperatures and densities in stellar cores the rates of fusion reactions are notoriously slow. For example, at solar core temperature $(T \sim 15MK)$ and density $(\sim 120\,g/cm^3)$, the energy release rate is only ~ 0.1 microwatt/cm^3—millions of times less than the rate of energy release of ordinary candela and thousands of times less than the rate at which a human body generates heat. Thus, reproduction of stellar core conditions in a lab for nuclear fusion power production is completely impractical. Because nuclear reaction rates strongly depend on temperature $(\sim exp(-E/kT))$, then in order to achieve reasonable rates of energy production in terrestrial fusion reactors 10–100 times higher temperatures (compared to stellar interiors) are required T~0.1–1.0 GK.

Criteria and Candidates for Terrestrial Reactions

In man-made fusion, the primary fuel is not constrained to be protons and higher temperatures can be used, so reactions with larger cross-sections are chosen. This implies a lower Lawson criterion, and therefore less startup effort. Another concern is the production of neutrons, which activate the reactor structure radiologically, but also have the advantages of allowing volumetric extraction of the fusion energy and tritium breeding. Reactions that release no neutrons are referred to as *aneutronic*.

In order to be useful as a source of energy, a fusion reaction must satisfy several criteria. It must:

- Be exothermic: This may be obvious, but it limits the reactants to the low Z (number of protons) side of the curve of binding energy. It also makes helium-4 the most common product because of its extraordinarily tight binding, although He-3 and H-3 also show up;
- Involve low Z nuclei: This is because the electrostatic repulsion must be overcome before the nuclei are close enough to fuse;
- Have two reactants: At anything less than stellar densities, three body collisions are too improbable. It should be noted that in inertial confinement, both stellar densities and

temperatures are exceeded to compensate for the shortcomings of the third parameter of the Lawson criterion, ICF's very short confinement time;

have two or more products: This allows simultaneous conservation of energy and momentum without relying on the electromagnetic force;

conserve both protons and neutrons: The cross sections for the weak interaction are too small.

Few reactions meet these criteria. The following are those with the largest cross sections:

(1)	D	+	T	→		^{4}He	(3.5 MeV)	+		n	(14.1 MeV)				
(2i)	D	+	D	→		T	(1.01 MeV)	+		p	(3.02 MeV)				*50%*
(2ii)				→		^{3}He	(0.82 MeV)	+		n	(2.45 MeV)				*50%*
(3)	D	+	^{3}He	→		^{4}He	(3.6 MeV)	+		p	(14.7 MeV)				
(4)	T	+	T	→		^{4}He		+	2	n	+ 11.3 MeV				
(5)	^{3}He	+	^{3}He	→		^{4}He		+	2	p	+ 12.9 MeV				
(6i)	^{3}He	+	T	→		^{4}He		+		p		+	n	+ 12.1 MeV	*51%*
(6ii)				→		^{4}He	(4.8 MeV)	+		D	(9.5 MeV)				*43%*
(6iii)				→		^{4}He	(0.5 MeV)	+		n	(1.9 MeV)	+	p	(11.9 MeV)	*6%*
(7i)	D	+	^{6}Li	→	2	^{4}He	+ 22.4 MeV								*__%*
(7ii)				→		^{3}He		+		^{4}He		+	n	+ 2.56 MeV	*__%*
(7iii)				→		^{7}Li		+		p	+ 5.0 MeV				*__%*
(7iv)				→		^{7}Be		+		n	+ 3.4 MeV				*__%*
(8)	p	+	^{6}Li	→		^{4}He	(1.7 MeV)	+		^{3}He	(2.3 MeV)				
(9)	^{3}He	+	^{6}Li	→	2	^{4}He		+		p	+ 16.9 MeV				
(10)	p	+	^{11}B	→	3	^{4}He	+	8.7 MeV							

Note: p (protium), D (deuterium), and T (tritium) are shorthand notation for the main three isotopes of hydrogen.

For reactions with two products, the energy is divided between them in inverse proportion to their masses, as shown. In most reactions with three products, the distribution of energy varies. For reactions that can result in more than one set of products, the branching ratios are given.

Some reaction candidates can be eliminated at once. The D-^{6}Li reaction has no advantage compared to p-^{11}B because it is roughly as difficult to burn but produces substantially more neutrons through D-D side reactions. There is also a p-^{7}Li reaction, but the cross section is far too low, except possibly when $T_i > 1$ MeV, but at such high temperatures an endothermic, direct neutron-producing reaction also becomes very significant. Finally there is also a p-^{9}Be reaction, which is not only difficult to burn, but ^{9}Be can be easily induced to split into two alphas and a neutron.

In addition to the fusion reactions, the following reactions with neutrons are important in order to "breed" tritium in "dry" fusion bombs and some proposed fusion reactors:

$$n + {}^6Li \rightarrow T + {}^4He$$
$$n + {}^7Li \rightarrow T + {}^4He + n$$

To evaluate the usefulness of these reactions, in addition to the reactants, the products, and the energy released, one needs to know something about the cross section. Any given fusion device will have a maximum plasma pressure that it can sustain, and an economical device will always operate near this maximum. Given this pressure, the largest fusion output is obtained when the temperature is chosen so that $<\sigma v>/T^2$ is a maximum. This is also the temperature at which the value of the triple product $nT\tau$ required for ignition is a minimum (a plasma is "ignited" if the fusion reactions produce enough power to maintain the temperature without external heating). This optimum temperature and the value of $<\sigma v>/T^2$ at that temperature is given for a few of these reactions in the following table.

fuel	T [keV]	$<\sigma v>/T^2$ [m^3/s/keV^2]
D-T	13.6	1.24×10^{-24}
D-D	15	1.28×10^{-26}
D-^{3}He	58	2.24×10^{-26}
p-^{6}Li	66	1.46×10^{-27}
p-^{11}B	123	3.01×10^{-27}

Note that many of the reactions form chains. For instance, a reactor fueled with T and ^{3}He will create some D, which is then possible to use in the D + ^{3}He reaction if the energies are "right." An elegant idea is to combine the reactions and. The ^{3}He from reaction can react with ^{6}Li in reaction before completely thermalizing. This produces an energetic proton which in turn undergoes reaction before thermalizing. A detailed analysis shows that this idea will not really work well, but it is a good example of a case where the usual assumption of a Maxwellian plasma is not appropriate.

Neutronicity, Confinement Requirement, and Power Density

Any of the reactions above can in principle be the basis of fusion power production. In addition to the temperature and cross section discussed above, we must consider the total energy of the fusion products E_{fus}, the energy of the charged fusion products E_{ch}, and the atomic number Z of the non-hydrogenic reactant.

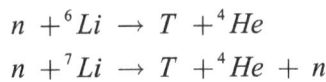

The only fusion reactions thus far produced by humans to achieve ignition are those which have been created in hydrogen bombs; the first of which, shot Ivy Mike, is shown here.

Specification of the D-D reaction entails some difficulties, though. To begin with, one must average over the two branches (2) and (3). More difficult is to decide how to treat the T and ^3He products. T burns so well in a deuterium plasma that it is almost impossible to extract from the plasma. The D-^3He reaction is optimized at a much higher temperature, so the burnup at the optimum D-D temperature may be low, so it seems reasonable to assume the T but not the ^3He gets burned up and adds its energy to the net reaction. Thus we will count the DD fusion energy as $E_{fus} = (4.03+17.6+3.27)/2 = 12.5$ MeV and the energy in charged particles as $E_{ch} = (4.03+3.5+0.82)/2 = 4.2$ MeV.

Another unique aspect of the D-D reaction is that there is only one reactant, which must be taken into account when calculating the reaction rate.

With this choice, we tabulate parameters for four of the most important reactions.

fuel	Z	E_{fus} [MeV]	E_{ch} [MeV]	neutronicity
D-T	1	17.6	3.5	0.80
D-D	1	12.5	4.2	0.66
D-^3He	2	18.3	18.3	~0.05
p-^{11}B	5	8.7	8.7	~0.001

The last column is the neutronicity of the reaction, the fraction of the fusion energy released as neutrons. This is an important indicator of the magnitude of the problems associated with neutrons like radiation damage, biological shielding, remote handling, and safety. For the first two reactions it is calculated as $(E_{fus}-E_{ch})/E_{fus}$. For the last two reactions, where this calculation would give zero, the values quoted are rough estimates based on side reactions that produce neutrons in a plasma in thermal equilibrium.

Of course, the reactants should also be mixed in the optimal proportions. This is the case when each reactant ion plus its associated electrons accounts for half the pressure. Assuming that the total pressure is fixed, this means that density of the non-hydrogenic ion is smaller than that of the hydrogenic ion by a factor $2/(Z+1)$. Therefore the rate for these reactions is reduced by the same factor, on top of any differences in the values of $\langle\sigma v\rangle/T^2$. On the other hand, because the D-D reaction has only one reactant, the rate is twice as high as if the fuel were divided between two hydrogenic species.

Thus there is a "penalty" of (2/(Z+1)) for non-hydrogenic fuels arising from the fact that they require more electrons, which take up pressure without participating in the fusion reaction (It is usually a good assumption that the electron temperature will be nearly equal to the ion temperature. Some authors, however, discuss the possibility that the electrons could be maintained substantially colder than the ions. In such a case, known as a "hot ion mode," the "penalty" would not apply. There is at the same time a "bonus" of a factor 2 for D-D due to the fact that each ion can react with any of the other ions, not just a fraction of them.

We can now compare these reactions in the following table:

fuel	$\langle\sigma v\rangle/T^2$	penalty/bonus	reactivity	Lawson criterion	power density
D-T	1.24×10^{-24}	1	1	1	1
D-D	1.28×10^{-26}	2	48	30	68
D-^{3}He	2.24×10^{-26}	2/3	83	16	80
p-^{11}B	3.01×10^{-27}	1/3	1240	500	2500

The maximum value of $\langle\sigma v\rangle/T^2$ is taken from a previous table. The "penalty/bonus" factor is that related to a non-hydrogenic reactant or a single-species reaction. The values in the column "reactivity" are found by dividing 1.24×10^{-24} by the product of the second and third columns. It indicates the factor by which the other reactions occur more slowly than the D-T reaction under comparable conditions. The column "Lawson criterion" weights these results with E_{ch} and gives an indication of how much more difficult it is to achieve ignition with these reactions, relative to the difficulty for the D-T reaction. The last column is labeled "power density" and weights the practical reactivity with E_{fus}. It indicates how much lower the fusion power density of the other reactions is compared to the D-T reaction and can be considered a measure of the economic potential.

Bremsstrahlung Losses in Quasineutral, Isotropic Plasmas

The ions undergoing fusion in many systems will essentially never occur alone but will be mixed with electrons that in aggregate neutralize the ions' bulk electrical charge and form a plasma. The electrons will generally have a temperature comparable to or greater than that of the ions, so they will collide with the ions and emit x-ray radiation of 10–30 keV energy (Bremsstrahlung). The sun and stars are opaque to x-rays, but essentially any terrestrial fusion reactor will be optically thin for x-rays of this energy range. X-rays are difficult to reflect but they are effectively absorbed (and converted into heat) in less than mm thickness of stainless steel (which is part of reactor shield). The ratio of fusion power produced to x-ray radiation lost to walls is an important figure of merit. This ratio is generally maximized at a much higher temperature than that which maximizes the power density. The following table shows the rough optimum temperature and the power ratio at that temperature for several reactions.

fuel	T_i (keV)	$P_{fusion}/P_{Bremsstrahlung}$
D-T	50	140
D-D	500	2.9
D-^{3}He	100	5.3
^{3}He-^{3}He	1000	0.72
p-^{6}Li	800	0.21
p-^{11}B	300	0.57

The actual ratios of fusion to Bremsstrahlung power will likely be significantly lower for several reasons. For one, the calculation assumes that the energy of the fusion products is transmitted completely to the fuel ions, which then lose energy to the electrons by collisions, which in turn lose energy by Bremsstrahlung. However because the fusion products move much faster than the fuel ions, they will give up a significant fraction of their energy directly to the electrons. Secondly, the plasma is assumed to be composed purely of fuel ions. In practice, there will be a significant proportion of impurity ions, which will lower the ratio. In particular, the fusion products themselves *must* remain in the plasma until they have given up their energy, and *will* remain some time after that in any proposed confinement scheme. Finally, all channels of energy loss other than Bremsstrahlung have been neglected. The last two factors are related. On theoretical and experimental grounds, particle and energy confinement seem to be closely related. In a confinement scheme that does a good job of retaining energy, fusion products will build up. If the fusion products are efficiently ejected, then energy confinement will be poor, too.

The temperatures maximizing the fusion power compared to the Bremsstrahlung are in every case higher than the temperature that maximizes the power density and minimizes the required value of the fusion triple product. This will not change the optimum operating point for D-T very much because the Bremsstrahlung fraction is low, but it will push the other fuels into regimes where the power density relative to D-T is even lower and the required confinement even more difficult to achieve. For D-D and D-^{3}He, Bremsstrahlung losses will be a serious, possibly prohibitive problem. For ^{3}He-^{3}He, p-^{6}Li and p-^{11}B the Bremsstrahlung losses appear to make a fusion reactor using these fuels with a quasineutral, anisotropic plasma impossible. Some ways out of this dilemma are considered—and rejected—in "Fundamental limitations on plasma fusion systems not in thermodynamic equilibrium" by Todd Rider. This limitation does not apply to non-neutral and anisotropic plasmas; however, these have their own challenges to contend with.

Thermonuclear Fusion

Thermonuclear fusion is actually quite similar to most chemical reactions in that two initial 'reactants' come into proximity to each other and join together to create a number of new products. Unlike chemical reactions, thermonuclear fusion is the fusion of the nuclei rather than atoms and molecules. Because of this difference there are different rules which must be obeyed and different problems which must be overcome.

The foremost problem fusion faces is that the two reactants are both nuclei – as such they are both positively charged bodies which we want to force together. This presents a problem as similar charges repel each other. It is particularly a problem because to fuse the nuclei, we need to bring the nuclei to within about 2×10^{-15} m of within each other. Only nuclei at extremely high temperatures have the required kinetic energy to overcome this potential barrier and so this is where the term thermonuclear fusion comes from.

Thermonuclear Fusion is a process which transforms matter into energy; the mass of the initial reactants is greater than the mass of the final products – and, as Einstein showed the world, the energy associated with a material is proportional to the product of the mass of the material and the square of the speed of light; $E = \gamma mc^2$. It is this mass deficit 1 which allows fusion to release energy.

Confinement

Stellar Thermonuclear Reactions

At normal temperatures the particles in a gas are neutral. But at temperatures above a few electron volts (eV), the individual particles tend to separate into their constituent elements (ions and electrons) and the gas gets transformed into a mixture of charged particles, that is, a plasma. About 99% of the matter making up the Universe consists of plasma - also called the "fourth state of matter". It is the main constituent of the sun and stars. The sun's interior temperature is 14 million degrees and here the reaction resulting from the fusion of hydrogen nuclei is responsible for most of the energy which reaches us on earth in the form of heat and light (and solar neutrinos).

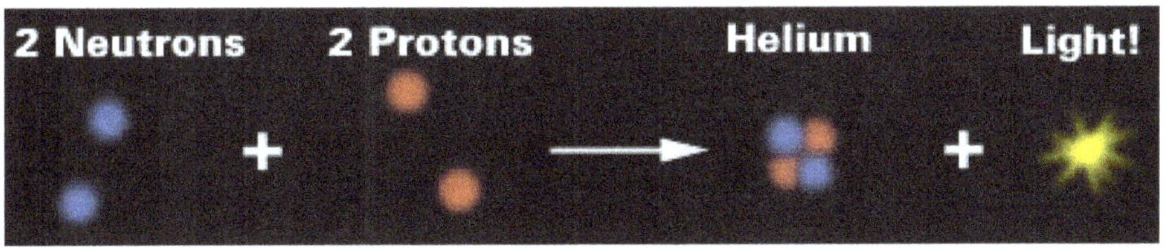

In the hotter or more massive stars, other reactions predominate. At around 15-20 million degrees of temperature, these reactions are based on the carbon cycle, in which carbon-12 acts as a catalyst to fuse 4 protons to a nucleus of He^4, two positrons, two neutrinos, and a gamma, producing 26.63 MeV of energy (5% of which associated with the neutrinos produced). Stellar evolution depends on the fusion energy and on gravitational energy. In a young star, consisting mostly of hydrogen atoms, gravitational energy is dominant, the star contracts, and its temperature and density increase until fusion reactions become significant enough for energy to be released. The gravitational and nuclear stages evolve sequentially at ever higher temperatures and densities and the burning nuclei become increasingly charged until they react into iron nuclei with maximum binding energy. At this point, the nuclear reactions begin to absorb energy rather than produce it.

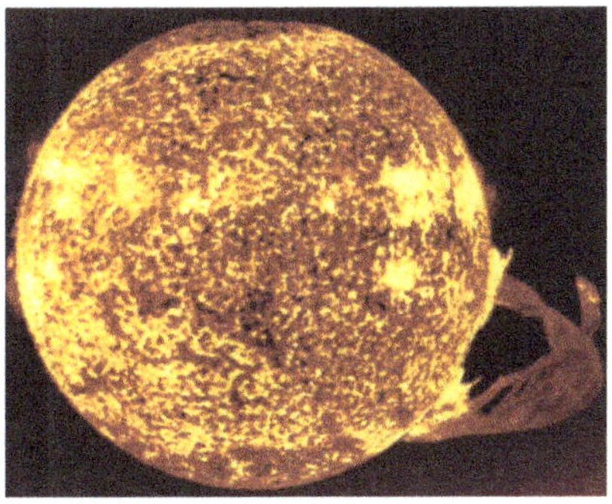

To obtain a fusion reaction, the hydrogen plasma has to be confined in a limited space. In the sun this situation is brought about by enormous gravitational forces. The fusion process in the sun occurs extremely slowly, which is why it has kept on shining for billions of years.

Magnetic Confinement of Plasmas

In magnetic confinement fusion the hot plasma is contained in a vacuum vessel, in which an appropriately configured external magnetic field and/or a field generated by a current induced in the plasma itself prevents the plasma from hitting the vessel walls.

Diverse magnetic configurations have been studied, for example, mirror configurations, in which the force lines of the magnetic field are open linearly, and toroidally symmetric configurations (e.g., stellarator and tokamak).

The concept that has given the best results so far in magnetic confinement fusion is the tokamak.

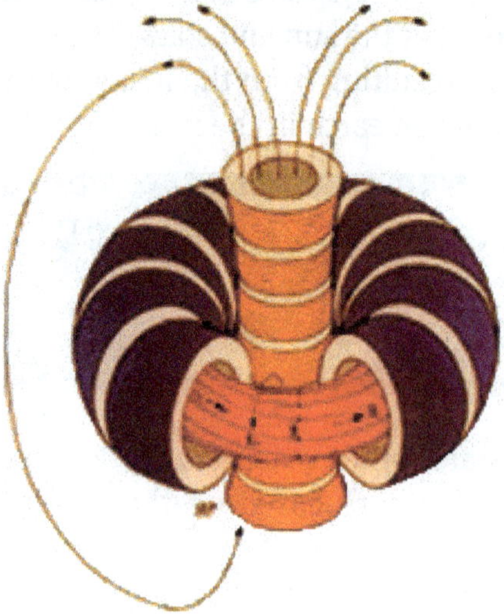

The tokamak is a toroidally shaped device characterised by a hollow vessel or chamber, forming the "doughnut", in which the plasma is confined by a magnetic field and bound to force field lines along a spiralling path.

This type of magnetic configuration is obtained by combining an intense toroidal magnetic field, produced by magnetic coils placed around the doughnut, with a poloidal magnetic field, obtained by externally inducing a current in the plasma. The poloidal current also helps to prevent the plasma particles from migrating towards the vessel walls.

The plasma particles spiral around the force field lines.

Another set of external magnetic coils is used to provide auxiliary magnetic fields that control the position of the plasma in the doughnut.

The tokamak configuration is particularly stable and allows the plasma to be confined for a long time.

Plasma Heating

As the plasma is an electric conductor it can be heated by inducing a current from the outside. The plasma in the doughnut forms the secondary circuit of a transformer whose primary circuit is external.

So, the induced current has two purposes: to generate the poloidal field and to heat the plasma to high temperatures ("4 current" in the figure below). This type of heating is called "Ohmic" or "resistive" and obeys Joule's law. Actually, it is similar to the heating that occurs in an electric lamp or heater.

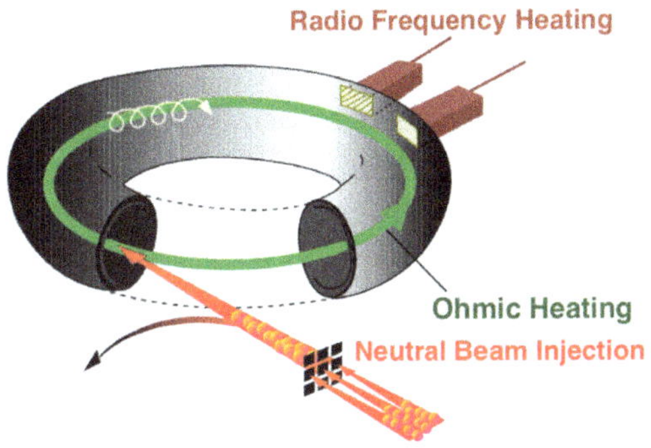

However, the effect of Ohmic heating is limited ("4 current" in figure) because the resistance of the plasma decreases as the temperature increases, so the maximum temperature that can be obtained in the plasma is only a few million degrees. If we want to reach the temperatures necessary for thermonuclear fusion, we have to use additional heating methods:

- Absorption of electromagnetic waves, which are injected into the plasma by means of wave-guides or antennas;
- Injection of neutral atoms at high kinetic energy, which cross the magnetic field, become ionised and transfer their energy to the plasma through collisions;
- Adiabatic compression of the plasma, obtained by moving the plasma to regions where the magnetic field is higher, which consequently increases the plasma temperature.

Inertial Confinement Fusion

Inertial Fusion apparatus: ABC plant

Inertial confinement is another line of fusion research that has been developed according to the Lawson criterion . Here, a series of micro-explosions is obtained through bombarding a small sphere containing a deuterium-tritium mixture by high-energy laser or ion beams. The electromagnetic energy of high-powered laser beams (or the kinetic energy of the accelerated particles) is uniformly transferred onto the sphere's surface, which evaporates and, according to the action-reaction principle, the fuel is compressed and heated. This gives a very high density plasma, but it is confined only for a very short time.

Advantages of Thermonuclear Fusion

Thermonuclear fusion presents an opportunity to provide the required energy in an environmentally friendly way. There are a number of reasons for fusion being environmentally friendly.

Firstly the reactants used can be extracted from water supplies, given that most of the surface of the Earth is covered in water, the supplies of hydrogen isotopes are abundant and should be easily accessible. The comparatively safe refining of this water would require a refining plant - however such a plant would have much less of an environmental impact than other means.

Secondly reactors in which the refined isotopes of hydrogen are placed will require the use of materials, however such uses are comparable to other reactors used in power stations and so whilst they must be considered, are not a major issue.

Also, the products of the reaction are simply helium, hydrogen or neutrons. Unlike products of other processes such as carbon dioxide or radioactive materials, the products of fusion have no particularly undesirable effects.

Disadvantages of Thermonuclear Fusion

The main disadvantage of fusion so far, is that nobody has been able to make a reactor that can consistently achieve breakeven. This jargon means the level at which the energy output of the reactor equals the energy input to the reactor. Obviously a reactor operating at breakeven would provide no more energy than we originally had and so any reactor which is to be useful must achieve a greater level of efficiency than the breakeven. This is the area in which fusion research is currently most concentrated on.

There are a number of issues facing nuclear fusion which must be overcome before fusion research can provide an alternative to current methods of generating electricity.

Controlled Fusion

Controlled fusion is the fusion of light atomic nuclei that occurs at high temperatures under controlled conditions and is accompanied by an energy release. Thermonuclear reactions proceed slowly because of the Coulomb repulsion of positively charged nuclei. Therefore, fusion occurs with an appreciable intensity only between light nuclei that have a small positive charge and only at high temperatures, where the kinetic energy of colliding nuclei is sufficient to overcome the Coulomb potential barrier.

Under natural conditions, thermonuclear reactions between hydrogen nuclei, or protons, occur deep inside stars, particularly in the interior regions of the sun, and thus serve as a constant energy source that governs the radiation of stars. The burning of hydrogen in stars occurs at a low rate, but the huge dimensions and densities of the stars ensure the continuous emission of tremendous energy fluxes for billions of years. Reactions between the heavy isotopes of hydrogen, deuterium ^2H and tritium ^3H, occur at an incomparably higher rate and result in the formation of strongly bound helium nuclei:

$$^2H + {}^2H \begin{cases} \nearrow {}^3He + n + 3.28 \text{ million electron volts (MeV)} \\ \searrow {}^3H + p\ 4.03 \text{MeV} \end{cases}$$
$$^2H + {}^3H \rightarrow {}^4He + n\ 17.6 \text{ MeV}$$

These reactions are of the greatest interest in controlled fusion. The second reaction, which is accompanied by a large energy release and occurs at a significant rate, is particularly interesting. Tritium is radioactive, with a half-life of 12.5 years, and is not found in nature. Consequently, to ensure the operation of a proposed fusion reactor that uses tritium as the fuel, the tritium must be bred. For this purpose, the reaction zone of the proposed system may be surrounded by a blanket of a light isotope of lithium, and the following breeding process would occur in that blanket:

$$^6Li + n \rightarrow {}^3H + {}^4He$$

The probability or effective cross section of thermonuclear reactions increases rapidly with temperature but, even under optimum conditions, remains incomparably smaller than the probability or effective cross section of atomic collisions. For this reason, fusion reactions must occur in a completely ionized plasma heated to a high temperature. In such a plasma, the ionization and excitation of atoms do not occur and, sooner or later, deuteron-deuteron or deuteron-triton collisions culminate in nuclear fusion.

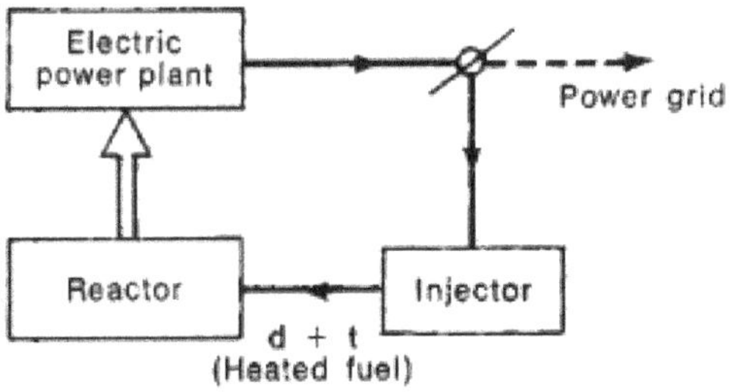

The specific power output of a fusion reactor is obtained by multiplying the number of nuclear reactions that occur each second per unit volume of the reaction zone of the reactor by the energy released in each reaction event.

The Lawson criterion. The use of the laws of conservation of energy and particle number makes it possible to elucidate some general requirements imposed on a fusion reactor, requirements that are independent of any particular engineering or design characteristics of the system in question. A schematic diagram of the operation of a reactor is provided in above figure. A device of arbitrary design contains a pure hydrogen plasma with a density n and a temperature T. Fuel, such as a mixture of equal parts of deuterium and tritium already heated to the required temperature, is injected into the reactor. Inside the reactor the injected particles collide with one another from time to time, and nuclear interactions occur. This is a useful process. At the same time, however, energy escapes from the reactor as a result of the electromagnetic radiation of the plasma, and some "hot," that is, high-energy, particles that have not undergone nuclear interactions escape from the reaction zone.

Let τ be the mean particle confinement time in the reactor; the meaning of the quantity τ is that, on the average, n/τ particles of each sign escape from 1 cm³ of the plasma in 1 sec. In steady-state operation the same number of particles (calculated per unit volume) must be injected into the reactor each second. To cover the energy losses, the fuel supplied must be fed into the reaction zone with an energy exceeding the energy of the flux of escaping particles. This additional energy must be compensated for by the fusion energy released in the reaction zone and by the partial recovery of electromagnetic radiation and corpuscular fluxes in the reactor walls and blanket.

Let us assume, for simplicity's sake, that the ratio of conversion into electrical energy is identical for the nuclear reaction products, electromagnetic radiation, and thermal particles and is equal to η. The quantity η is often called the efficiency. When the system operates in the steady state and the effective power output is zero, the energy balance equation for the reactor has the form

$$\eta(P_0+P_r+P_t) = P_r+P_t$$

where P_0 is the energy released as nuclear power, P_r is the power of the radiation flux, and P_t is the power of the escaping particle flux. When the left-hand side of the equation is made greater than the right-hand side, the reactor stops expending energy and begins operating as a thermonuclear electric power plant. In writing equation $\eta(P_0+P_r+P_t) = P_r+P_t$, it is assumed that all the recovered energy is returned without losses to the reactor through the injector, together with the flow of the heated fuel that is supplied. Since the quantities P_0, P_r, and P_t depend in a known way on the plasma temperature, we can easily calculate from the balance equation the product

$$n\tau = f(T)$$

Here, $f(T)$ for a given value of the efficiency η and the type of fuel selected is a well-defined function of temperature. Plots of $f(T)$ are given in above Figure for two values of η and for both deuteron-deuteron (d, d) and deuteron-triton (d, t) reactions. If the values of nτ attained in a given device lie above the curve of $f(T)$, the system will operate as an energy generator. When $\eta = 1/3$, the operation of the reactor in producing useful power corresponds under optimum conditions (the minimum of the curves in above figure) to the following condition, which is called the Lawson criterion: for (d, d) reactions, $n\tau \geq 10^{15}\ cm^{-3} \cdot sec$, and $T \sim 10^{90} K$; for (d, t) reactions, $n\tau \geq 0.5 \times 10^{14}\ cm^{-3} \cdot sec$, and $T \sim 2 \times 10^{80} K$.

Thus, even under optimum conditions and with very optimistic assumptions concerning the value of η temperatures of about $2\times10^{80} K$ must be attained for the case of greatest interest, namely, a reactor fueled by a mixture of equal parts of deuterium and tritium. In this case, confinement times of the order of seconds must be achieved for a plasma with a density of about $10^{14} cm^{-3}$. The reactor, of course, may produce useful energy at lower temperatures, but the energy must be "paid for" by higher values of nτ.

In short, the construction of a fusion reactor presupposes the creation of a plasma heated to temperatures of hundreds of millions of degrees and the maintenance of a plasma configuration for the time required for nuclear reactions to occur. Research in controlled fusion is following two paths—the development of quasi-steady-state systems, on the one hand, and the development of extremely high-speed devices, on the other.

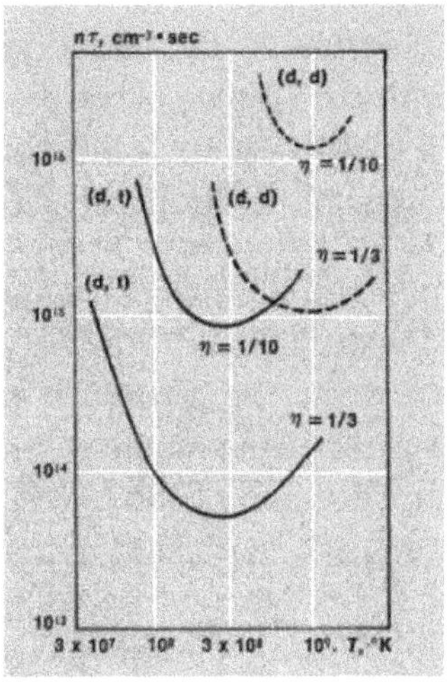

Controlled fusion with magnetic confinement. Let us first consider quasi-steady-state systems. An energy yield at the level of 10^5 kilowatts per cubic meter (kW/m^3) is achieved for (d, t) reactions when the plasma density is about 10^{15} cm^{-3} and the temperature is about $10^{8\circ}$K. This means that the size of a reactor with an output of 10^6–10^7 kW—the typical output of large present-day electric power plants—should be in the range 10–100 m^3, which is entirely acceptable. The main problem is how to confine the hot plasma in the reaction zone. The diffusive particle and heat fluxes at the cited values of n and T ate huge, and no material walls are suitable. A ground-breaking idea advanced in 1950 in the Soviet Union and the USA consists in the use of the principle of magnetic confinement of the plasma. When the charged particles that form the plasma are located in a magnetic field, they cannot move freely perpendicular to the field's lines of force. As a result, the coefficients of diffusion and thermal conductivity across the magnetic field decrease very rapidly with increasing field strength in the case of a stable plasma; for example, with fields of $\sim$ 10 gauss the coefficients are reduced by 14–15 orders of magnitude as opposed to their "unmagnetized" value for a plasma with the density and temperature indicated above. Thus, in principle, the use of a sufficiently strong magnetic field opens the way to the design of a fusion reactor.

There are three areas of research in the field of controlled fusion with magnetic confinement:

- open, or mirror-type, magnetic traps,
- closed magnetic systems, and
- pulsed machines.

In open traps, particles cannot easily escape, across the lines of force, from the reaction zone to the walls of the device. The particles escape either during "magnetized" diffusion—that is, very slowly—or by means of charge exchange with the molecules of the residual, that is, un-ionized gas. The escape of the plasma along the lines of force is likewise inhibited by regions of intensified magnetic field located at the open ends of the trap; such regions are called magnetic mirrors, or end plugs. The traps are usually filled with plasma by injecting plasmoids or individual high-energy particles. Additional plasma heating can be accomplished with the aid of adiabatic compression in a rising magnetic field.

In closed systems, such as Tokamaks and stellarators, the escape of particles to the walls of a toroidal device across a longitudinal magnetic field is also difficult and occurs as a result of magnetized diffusion and charge exchange. The plasma column in a Tokamak is heated in the initial stages by a ring current that flows through the column. As the temperature rises, however, Joule heating becomes increasingly less effective, since the resistance of the plasma decreases rapidly with increasing temperature. Methods of heating by a high-frequency electromagnetic field and by the injection of energy with the aid of fluxes of fast neutral particles are used to heat the plasma above 10^{7}°K.

In pulsed machines, such as the Z pinch and 8 pinch, plasma heating and plasma confinement are accomplished by strong short-period currents that flow through the plasma. As the current and magnetic pressure increase simultaneously, the plasma is squeezed away from the walls of the containment vessel. This effect ensures that the plasma is confined. The temperature increases as a result of Joule heating, adiabatic compression of the plasma column, and, apparently, turbulent processes associated with the development of a plasma instability.

The study of hot plasmas in high-frequency (HF) fields is an independent area of research. As the experiments of P. L. Kapitsa have shown, in hydrogen and helium a freely hovering plasma column with an electron temperature of about 10^{6}°K can be produced in HF fields at sufficiently high pressure. Such a system allows for the closing of the column into a ring and the superposition of an additional longitudinal magnetic field.

The successful operation of any of the devices listed above is possible only if the initial plasma structure is macroscopically stable and maintains a specified shape for the entire period of time necessary for the reaction to occur. Furthermore, microscopic instabilities must be suppressed in the plasma. When such instabilities arise and develop, the energy distribution of the particles ceases to be an equilibrium distribution, and the particle and heat fluxes across the lines of force increase sharply in comparison with their theoretical values. Since 1950 most research in magnetic systems has been directed at the stabilization of plasma configurations; this work still cannot be regarded as completed.

Ultrahigh-speed controlled fusion systems with inertia confinement. The difficulties associated with the magnetic confinement of a plasma can be obviated in principle if the nuclear fuel is burned for extremely short periods of time, during which the heated matter cannot disperse from the reaction zone. In accordance with the Lawson criterion, useful energy can be obtained with this method of burning only if the fuel has a very high density. To avoid a high-power thermonuclear explosion, very small amounts of fuel must be used. The initial thermonuclear fuel must be in the form of small pellets, 1–2 mm in diameter, prepared from a mixture of deuterium and tritium and injected into the reactor before each reaction cycle. The main problem consists in supplying the necessary energy to heat the fuel pellets. As of 1976, the solution of this problem lies in the use of laser beams or high-power electron beams. Research in controlled fusion with the use of laser heating was begun in 1964; the use of electron beams is in an early stage of study—thus far, relatively few electron-beam fusion experiments have been performed.

Estimates show that the energy W that must be supplied to sustain the operation of a reactor is expressed as

$$W \geq \frac{10^{8}}{\eta^{3}\alpha^{2}} \text{ joules (J)}$$

Here, η is a general expression for the efficiency of the device and α is the coefficient of target compression. As this equation shows, even with very optimistic assumptions regarding the possible value of η the value of W when $\alpha = 1$ is disproportionately large. Admissible values of Wean therefore be approached only in conjunction with a sharp increase in the density of the target (by a factor of approximately 10^4) in comparison with the initial density of a solid (d, t) target. Rapid heating of the target is accompanied by the vaporization of its surface layers and the reaction compression of its interior regions. If the power supplied is time-programmed in a specific manner, then, as calculations show, the coefficients of compression indicated can be attained. Another possibility is to program the radial distribution of the target density. In both cases, the energy required is reduced to 106J, which is technically feasible considering the rapid development of laser devices.

Difficulties and prospects. Research in controlled fusion encounters major difficulties that are both purely physical and technical in nature. The physical difficulties include the cited problem of the stability of a hot plasma placed in a magnetic trap. It is true that the use of strong magnetic fields with a special configuration suppresses the particle fluxes escaping from the reaction zone and, in a number of cases, makes it possible to obtain sufficiently stable plasma formations. With n of about 10^{15} cm^{-3}, T of about 10^{8o}K, and a possible reactor size of 10–100 m^3, electromagnetic radiation freely escapes from the plasma. However, for a purely hydrogen plasma, the energy losses are determined solely by electron bremsstrahlung and, in the case of (d, t) reactions, are compensated for by the energy released as nuclear power at temperatures above 4×10^{7o}K.

The second fundamental difficulty is connected with the problem of impurities. Even a small impurity of foreign atoms with a high atomic weight, which are in a highly ionized state at the temperatures considered, leads to a sharp increase in the intensity of the continuous spectrum, to the appearance of a line spectrum, and to an increase in energy losses to a level above the acceptable level. Extraordinary efforts, such as the continuous improvement of evacuation equipment, the use of refractory and nonsputtering metals as the orifice material, and the use of special devices to trap foreign atoms, are required to minimize the amount of impurities in the plasma. More accurately, the "lethal" concentration that makes it impossible for thermonuclear reactions to occur is, for example, a few tenths of 1 percent for a tungsten or molybdenum impurity.

The parameters achieved in various machines as of mid-1976 are shown in below figure in a plot of $n\tau$ against T. Tokamaks and laser systems come the closest to the region where the Lawson criterion is met, and a self-sustaining thermonuclear reaction can occur. It would be erroneous, however, to make categorical conclusions, on the basis of the available data, as to the type of device that will be used as a future thermonuclear reactor. The development of this field of technical physics is proceeding too rapidly, and many estimates will change over the next decade.

The tremendous importance attached to research in controlled fusion is explained by a number of factors. The growing pollution of the environment urgently requires the conversion of the planet's industrial production to a closed cycle with a minimum of waste. However, such a reorganization of industry inevitably entails a sharp increase in energy consumption. Meanwhile, the resources of fossil fuels are limited and, at the current rate of development of power engineering, will be exhausted in the next few decades—as in the case of petroleum and gaseous fuels—or centuries—as in the case of coal. The best alternative, of course, would be the use of solar energy, but the lower power density of incident solar radiation greatly complicates the efficient solution of this problem. The conversion to the use of nuclear fission reactors.

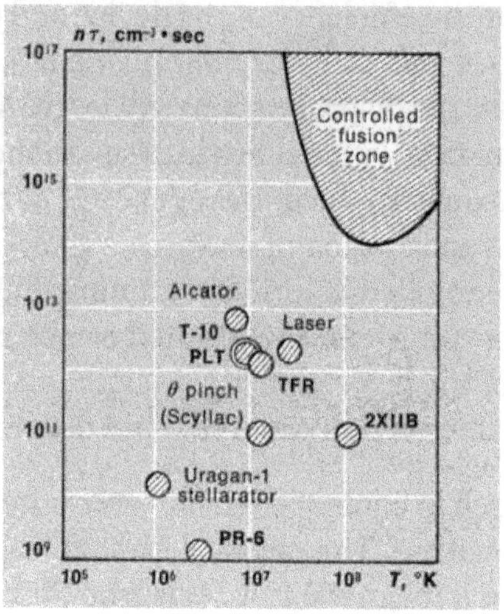

In above figure Parameters attained by mid-1976 in various machines for studying the problem of controlled fusion: (T-10) a Tokamak machine at the I. V. Kurchatov Institute of Atomic Energy, USSR; (PLT) a Tokamak machine at the Princeton Plasma Physics Laboratory, USA; (Alcator) a Tokamak machine at the Massachusetts Institute of Technology, USA; (TFR) a Tokamak machine at Fontanay-aux-Roses, France; (PR-6) an open trap at the I. V. Kurchatov Institute of Atomic Energy, USSR; (2XIIB) an open trap at the Lawrence Livermore Laboratory, USA; (o pinch [Scyllac]) a machine at the Los Alamos Scientific Laboratory, USA; (Uragan-1 stellarator) a machine at the Ukrainian Physlcotechnical Institute, USSR; (Laser) pulsed laser-heated systems, USSR and USA

on a global scale raises complex problems of the burial of tremendous amounts of radioactive wastes; the alternative is to eject the radioactive wastes into space. According to available estimates, the radiation danger of controlled fusion installations should be three orders of magnitude less than that of fission reactors. In the future, the best solution would be a combination of solar energy and controlled fusion.

Fusion Reaction

Fusion reactions constitute the fundamental energy source of stars, including the Sun. The evolution of stars can be viewed as a passage through various stages as thermonuclear reactions and nucleosynthesis cause compositional changes over long time spans. Hydrogen (H) "burning" initiates the fusion energy source of stars and leads to the formation of helium (He). Generation of fusion energy for practical use also relies on fusion reactions between the lightest elements that burn to form helium. In fact, the heavy isotopes of hydrogen—deuterium (D) and tritium (T)—react more efficiently with each other, and, when they do undergo fusion, they yield more energy per reaction than do two hydrogen nuclei. (The hydrogen nucleus consists of a single proton. The deuterium nucleus has one proton and one neutron, while tritium has one proton and two neutrons.)

Fusion reactions between light elements, like fission reactions that split heavy elements, release energy because of a key feature of nuclear matter called the binding energy, which can be released through fusion or fission. The binding energy of the nucleus is a measure of the efficiency with which its constituent nucleons are bound together. Take, for example, an element with Z protons and N neutrons in its nucleus. The element's atomic weight A is $Z + N$, and its atomic number is Z. The binding energy B is the energy associated with the mass difference between the Z protons and N neutrons considered separately and the nucleons bound together $(Z + N)$ in a nucleus of mass M. The formula is

$$B = (Zm_p + Nm_n - M)c^2,$$

where m_p and m_n are the proton and neutron masses and c is the speed of light. It has been determined experimentally that the binding energy per nucleon is a maximum of about 1.4 10^{-12} joule at an atomic mass number of approximately 60—that is, approximately the atomic mass number of iron. Accordingly, the fusion of elements lighter than iron or the splitting of heavier ones generally leads to a net release of energy.

Types of Fusion Reactions

Fusion reactions are of two basic types: (1) those that preserve the number of protons and neutrons and (2) those that involve a conversion between protons and neutrons. Reactions of the first type are most important for practical fusion energy production, whereas those of the second type are crucial to the initiation of star burning. An arbitrary element is indicated by the notation *AZX*, where Z is the charge of the nucleus and A is the atomic weight. An important fusion reaction for practical energy generation is that between deuterium and tritium (the D-T fusion reaction). It produces helium (He) and a neutron (n) and is written:

$$D + T \rightarrow He + n.$$

To the left of the arrow (before the reaction) there are two protons and three neutrons. The same is true on the right.

The other reaction, that which initiates star burning, involves the fusion of two hydrogen nuclei to form deuterium (the H-H fusion reaction):

$$\mathrm{H} + \mathrm{H} \rightarrow \mathrm{D} + \beta^+ + \nu,$$

where β^+ represents a positron and ν stands for a neutrino. Before the reaction there are two hydrogen nuclei (that is, two protons). Afterward there are one proton and one neutron (bound together as the nucleus of deuterium) plus a positron and a neutrino (produced as a consequence of the conversion of one proton to a neutron).

Both of these fusion reactions are exoergic and so yield energy. The German-born physicist Hans Bethe proposed in the 1930s that the H-H fusion reaction could occur with a net release of energy and provide, along with subsequent reactions, the fundamental energy source sustaining the stars. However, practical energy generation requires the D-T reaction for two reasons: first, the rate of reactions between deuterium and tritium is much higher than that between protons; second, the net energy release from the D-T reaction is 40 times greater than that from the H-H reaction.

Energy Released in Fusion Reactions

Energy is released in a nuclear reaction if the total mass of the resultant particles is less than the mass of the initial reactants. To illustrate, suppose two nuclei, labeled X and a, react to form two other nuclei, Y and b, denoted

$$X+a\rightarrow Y+b.$$

The particles a and b are often nucleons, either protons or neutrons, but in general can be any nuclei. Assuming that none of the particles is internally excited (i.e., each is in its ground state), the energy quantity called the Q-value for this reaction is defined as

$$Q= (m_x+m_a-m_b-m_y)c^2,$$

where the m-letters refer to the mass of each particle and c is the speed of light. When the energy value Q is positive, the reaction is exoergic; when Q is negative, the reaction is endoergic (i.e., absorbs energy). When both the total proton number and the total neutron number are preserved before and after the reaction (as in D-T reactions), then the Q-value can be expressed in terms of the binding energy B of each particle as

$$Q=B_y+B_b-B_x-B_a.$$

The D-T fusion reaction has a positive Q-value of 2.8×10^{-12} joule. The H-H fusion reaction is also exoergic, with a Q-value of 6.7×10^{-14} joule. To develop a sense for these figures, one might consider that one metric ton (1,000 kg, or almost 2,205 pounds) of deuterium would contain roughly 3×10^{32} atoms. If one ton of deuterium were to be consumed through the fusion reaction with tritium, the energy released would be 8.4×10^{20} joules. This can be compared with the energy content of one ton of coal—namely, 2.9×10^{10} joules. In other words, one ton of deuterium has the energy equivalent of approximately 29 billion tons of coal.

Rate and Yield of Fusion Reactions

The energy yield of a reaction between nuclei and the rate of such reactions are both important. These quantities have a profound influence in scientific areas such as nuclear astrophysics and the potential for nuclear production of electrical energy.

When a particle of one type passes through a collection of particles of the same or different type, there is a measurable chance that the particles will interact. The particles may interact in many ways, such as simply scattering, which means that they change direction and exchange energy, or they may undergo a nuclear fusion reaction. The measure of the likelihood that particles will interact is called the cross section, and the magnitude of the cross section depends on the type of interaction and the state and energy of the particles. The product of the cross section and the atomic density of the target particle is called the macroscopic cross section. The inverse of the macroscopic cross section is particularly noteworthy as it gives the mean distance an incident particle will travel before interacting with a target particle; this inverse measure is called the mean free path. Cross sections are measured by producing a beam of one particle at a given energy, allowing the beam to interact with a (usually thin) target made of the same or a different material, and measuring deflections or reaction products. In this way it is possible to determine the relative likelihood of one type of fusion reaction versus another, as well as the optimal conditions for a particular reaction.

The cross sections of fusion reactions can be measured experimentally or calculated theoretically, and they have been determined for many reactions over a wide range of particle energies. They are well known for practical fusion energy applications and are reasonably well known, though with gaps, for stellar evolution. Fusion reactions between nuclei, each with a positive charge of one or more, are the most important for both practical applications and the nucleosynthesis of the light elements in the burning stages of stars. Yet, it is well known that two positively charged nuclei repel each other electrostatically—i.e., they experience a repulsive force inversely proportional to the square of the distance separating them. This repulsion is called the Coulomb barrier . It is highly unlikely that two positive nuclei will approach each other closely enough to undergo a fusion reaction unless they have sufficient energy to overcome the Coulomb barrier. As a result, the cross section for fusion reactions between charged particles is very small unless the energy of the particles is high, at least 10^4electron volts (1 eV ≅ 1.602 × 10^{-19} joule) and often more than 10^5 or 10^6 eV. This explains why the centre of a star must be hot for the fuel to burn and why fuel for practical fusion energy systems must be heated to at least 50,000,000 kelvins (K; 90,000,000 °F). Only then will a reasonable fusion reaction rate and power output be achieved.

The phenomenon of the Coulomb barrier also explains a fundamental difference between energy generation by nuclear fusion and nuclear fission. While fission of heavy elements can be induced by either protons or neutrons, generation of fission energy for practical applications is dependent on neutrons to induce fission reactions in uranium or plutonium. Having no electric charge, the neutron is free to enter the nucleus even if its energy corresponds to room temperature. Fusion energy, relying as it does on the fusion reaction between light nuclei, occurs only when the particles are sufficiently energetic to overcome the Coulomb repulsive force. This requires the production and heating of the gaseous reactants to the high temperature state known as the plasma state.

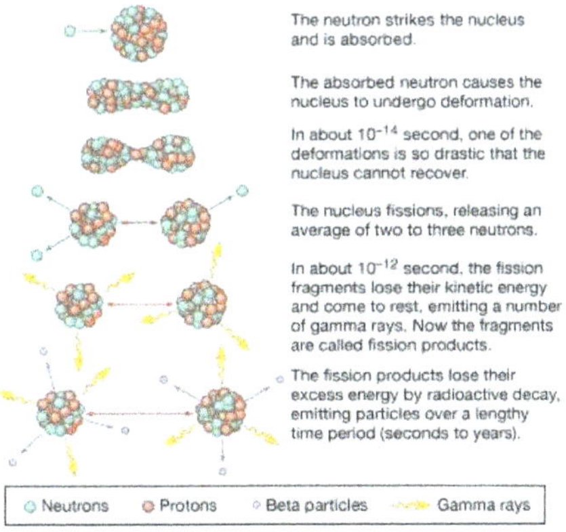
The neutron strikes the nucleus and is absorbed.
The absorbed neutron causes the nucleus to undergo deformation.
In about 10^{-14} second, one of the deformations is so drastic that the nucleus cannot recover.
The nucleus fissions, releasing an average of two to three neutrons.
In about 10^{-12} second, the fission fragments lose their kinetic energy and come to rest, emitting a number of gamma rays. Now the fragments are called fission products.
The fission products lose their excess energy by radioactive decay, emitting particles over a lengthy time period (seconds to years).
Neutrons
Protons
Beta particles
Gamma rays

Sequence of events in the fission of a uranium nucleus by a neutron.

Plasma State

Typically, a plasma is a gas that has had some substantial portion of its constituent atoms or molecules ionized by the dissociation of one or more of their electrons. These free electrons enable plasmas to conduct electric charges, and a plasma is the only state of matter in which thermonuclear reactions can occur in a self-sustaining manner. Astrophysics and magnetic fusion research, among other fields, require extensive knowledge of how gases behave in the plasma state. The stars, the solar wind, and much of interstellar space are examples where the matter present is in the plasma state. Very high-temperature plasmas are fully ionized gases, which means that the ratio of neutral gas atoms to charged particles is small. For example, the ionization energy of hydrogen is 13.6 eV, while the average energy of a hydrogen ion in a plasma at 50,000,000 K is 6,462 eV. Thus, essentially all of the hydrogen in this plasma would be ionized.

A reaction-rate parameter more appropriate to the plasma state is obtained by accounting for the fact that the particles in a plasma, as in any gas, have a distribution of energies. That is to say, not all particles have the same energy. In simple plasmas this energy distribution is given by the Maxwell-Boltzmann distribution law, and the temperature of the gas or plasma is, within a proportionality constant, two-thirds of the average particle energy; i.e., the relationship between the average energy *E* and temperature *T* is $3nkT$, where *k* is the Boltzmann constant, 8.62 × 10^{-5} eV per kelvin. The intensity of nuclear fusion reactions in a plasma is derived by averaging the product of the particles' speed and their cross sections over a distribution of speeds corresponding to a Maxwell-Boltzmann distribution. The cross section for the reaction depends on the energy or speed of the particles. The averaging process yields a function for a given reaction that depends only on the temperature and can be denoted $f(T)$. The rate of energy released (i.e., the power released) in a reaction between two species, *a* and *b*, is

$$P_{ab} = n_a n_b f_{ab}(T) U_{ab},$$

where n_a and n_b are the density of species *a* and *b* in the plasma, respectively, and U_{ab} is the energy released each time *a* and *b* undergo a fusion reaction. The parameter *Pab* properly takes into account both the rate of a given reaction and the energy yield per reaction.

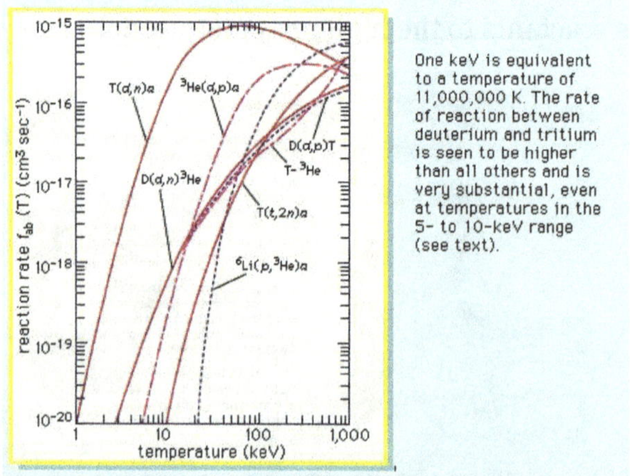

The reaction rate as a function of plasma temperature, expressed in kiloelectron volts (keV; 1 keV is equivalent to a temperature of 11,000,000 K). The rate of reaction between deuterium and tritium is seen to be higher than all others and is very substantial, even at temperatures in the 5-to-10-keV range.

Fusion Reactions in Stars

Fusion reactions are the primary energy source of stars and the mechanism for the nucleosynthesis of the light elements. In the late 1930s Hans Bethe first recognized that the fusion of hydrogen nuclei to form deuterium is exoergic (i.e., there is a net release of energy) and, together with subsequent nuclear reactions, leads to the synthesis of helium. The formation of helium is the main source of energy emitted by normal stars, such as the Sun, where the burning-core plasma has a temperature of less than 15,000,000 K. However, because the gas from which a star is formed often contains some heavier elements, notably carbon (C) and nitrogen (N), it is important to include nuclear reactions between protons and these nuclei. The reaction chain between protons that ultimately leads to helium is the proton-proton cycle. When protons also induce the burning of carbon and nitrogen, the CN cycle must be considered; and, when oxygen (O) is included, still another alternative scheme, the CNO bi-cycle, must be accounted for.

The proton-proton nuclear fusion cycle in a star containing only hydrogen begins with the reaction

$$H + H \rightarrow \mathrm{D} + \beta^{+} + \nu; Q = 1.44 \text{ MeV},$$

where the Q-value assumes annihilation of the positron by an electron. The deuterium could react with other deuterium nuclei, but, because there is so much hydrogen, the D/H ratio is held to very low values, typically 10^{-18}. Thus, the next step is

$$\mathrm{H} + \mathrm{D} \rightarrow {}^{3}\mathrm{He} + \gamma; Q = 5.49 \text{ MeV},$$

where γ indicates that gamma rays carry off some of the energy yield. The burning of the helium-3 isotope then gives rise to ordinary helium and hydrogen via the last step in the chain:

$${}^{3}\mathrm{He} + {}^{3}\mathrm{He} \rightarrow {}^{4}\mathrm{He} + 2(H); Q = 12.86 \text{ MeV}.$$

At equilibrium, helium-3 burns predominantly by reactions with itself because its reaction rate with hydrogen is small, while burning with deuterium is negligible due to the very low deuterium concentration. Once helium-4 builds up, reactions with helium-3 can lead to the production of still-heavier elements, including beryllium-7, beryllium-8, lithium-7, and boron-8, if the temperature is greater than about 10,000,000 K.

The stages of stellar evolution are the result of compositional changes over very long periods. The size of a star, on the other hand, is determined by a balance between the pressure exerted by the hot plasma and the gravitational force of the star's mass. The energy of the burning core is transported toward the surface of the star, where it is radiated at an effective temperature. The effective temperature of the Sun's surface is about 6,000 K, and significant amounts of radiation in the visible and infrared wavelength ranges are emitted.

Fusion Reactions for Controlled Power Generation

Reactions between deuterium and tritium are the most important fusion reactions for controlled power generation because the cross sections for their occurrence are high, the practical plasma temperatures required for net energy release are moderate, and the energy yield of the reactions are high—17.58 MeV for the basic D-T fusion reaction.

It should be noted that any plasma containing deuterium automatically produces some tritium and helium-3 from reactions of deuterium with other deuterium ions. Other fusion reactions involving elements with an atomic number above 2 can be used, but only with much greater difficulty. This is because the Coulomb barrier increases with increasing charge of the nuclei, leading to the requirement that the plasma temperature exceed 1,000,000,000 K if a significant rate is to be achieved. Some of the more interesting reactions are:

(1) $H + {}^{11}B \rightarrow 3({}^{4}He)$; Q = 8.68 MeV;

(2) $H + {}^{6}Li \rightarrow {}^{3}He + {}^{4}He$; Q = 4.023 MeV;

(3) ${}^{3}He + {}^{6}Li \rightarrow H + 2({}^{4}He)$; Q = 16.88 MeV; and

(4) ${}^{3}He + {}^{6}Li \rightarrow D + {}^{7}Be$; Q = 0.113 MeV.

Reaction (2) converts lithium-6 to helium-3 and ordinary helium. Interestingly, if reaction (2) is followed by reaction (3), then a proton will again be produced and be available to induce reaction (2), thereby propagating the process. Unfortunately, it appears that reaction (4) is 10 times more likely to occur than reaction (3).

Some Important Fusion Reactions

Main Controlled Fusion Fuels

First, we consider the reactions between the hydrogen isotopes deuterium and tritium, which are most important for controlled fusion research. Due to Z = 1, these hydrogen reactions have relatively small values of $\in_G$ and hence relatively large tunnel penetrability. They also have a relatively large S.

$$D + T \rightarrow \alpha\ (3.5\,\text{MeV}) + n\ (14.1\,\text{MeV})$$

The DT reaction has the largest cross-section, which reaches its maximum (about 5 barn) at the relatively modest energy of 64 keV. Its $Q_{DT} = 17.6\ MeV$ is the largest of this family of reactions. It is to be observed that the cross section of this reaction is characterized by a broad resonance for the formation of the compound He nucleus at $\in \cong 64\ keV$. Therefore, the astrophysical factor S exhibits a large variation in the energy interval of interest.

$$D + D \rightarrow T\ (1.01\ MeV) + p\,(3.03\ MeV),$$

$$D + D \rightarrow {}^{3}He\,(0.82\ MeV) + n\ (2.45\ MeV)$$

The DD reactions are nearly equiprobable. In the 10–100 keV energy interval, the cross sections for each of them are about 100 times smaller than for DT. The reaction $D(d, \gamma)^4$, He instead has cross section about 10,000 times smaller than that of 1.40 and 1.41.

$$T + T \rightarrow \alpha + 2n + 11.3\,\text{MeV},$$

The TT reaction has cross section comparable to that of DD. Notice that since the reaction has three products, the energies associated to each of them are not uniquely determined by conservation laws.

Advanced Fusion Fuels

Next, we consider reactions between hydrogen isotopes and light nuclei (Helium, Lithium, Boron). In the context of controlled fusion research mixtures of hydrogen and such elements are called advanced fusion fuels. For this group of reactions the Gamow energy is higher than for the previous group, leading to smaller cross-sections at relatively low energy. At high energy the cross sections are intermediate between that of DD and that of DT.

The proton–boron reaction

$$p + {}^{11}B \rightarrow 3\alpha + 8.6\,\text{MeV},$$

is particularly interesting, because it does not involve any radioactive fuel, and only releases charged particles. Its cross section exhibits a very narrow resonance at $\in = 148\ keV$, where the S factor peaks at 3500 MeV·barn and a broader resonance at $\in = 580\,\text{keV}$, where $S \approx 380\,\text{MeV}$ ·barn.

The D^3 He reaction also does not involve radioactive fuel and does not release neutrons, but a D^3He fuel would anyhow produce tritium and emit neutrons due to unavoidable DD reactions.

p–p Cycle

Reactions involved in the p–p cycle, the main source of energy in the Sun, are of fundamental importance in astrophysics. The first two reactions of the cycle, the pp reaction and the pD reaction have the lowest Gamow energy $\in_G$ of all fusion reactions, but their cross sections are much smaller than those of the previous reactions. Indeed, the pp reaction involves a low probability beta-decay, resulting in a value of S about 25 orders of magnitude smaller than that of the DT reaction. The pD reaction involves an electromagnetic transition, which is much more probable than pp, but still much less probable than reactions $D+T \rightarrow \alpha\ (3.5\,\text{MeV}) + n\ (14.1\,\text{MeV})$, $p + {}^{11}B \rightarrow 3\alpha + 8.6\,\text{MeV}$, based on strong interaction.

CNO Cycle

The reactions of the CNO cycle, the other main cycle responsible for energy production and hydrogen burning in stars. Here the S factors are not very small, but the Gamow energy takes values close to 40 MeV, thus resulting in cross sections smaller than those of the p–p cycle at relatively low temperatures. Indeed the p–p chain dominates in the Sun, which has central temperature of 1.3 keV . The CNO cycle, instead, prevails over the p–p cycle at temperatures larger than about 1.5 keV.

CC Reactions

Finally, Table lists data for the reactions between ${}^{12}C$ nuclei. Such nuclei are the main constituents of some white dwarfs. It is seen that the S factor is very large, but even at an energy of 100 keV the cross section is below $10^{-100}\ cm^2$, due to the extremely high Coulomb barrier. We shall see in Section that CC reactions become in fact possible in white dwarfs at densities above $10^9\ g/cm^3$.

Maxwell-averaged Fusion Reactivities

The effectiveness of a fusion fuel is characterized by its reactivity $\langle \sigma v \rangle$. Both in controlled fusion and in astrophysics we usually deal with mixtures of nuclei of different species, in thermal equilibrium, characterized by Maxwellian velocity distributions

$$f_j(v_j) = \left(\frac{mj}{2\pi k_B T}\right)^{3/2} \exp\left(-\frac{m_j v_j^2}{2k_B T}\right),$$

where the subscript j labels the species, T is the temperature and $_B$ is Boltzmann constant. The expression for the average reactivity 1.10 can now be written as

$$\langle \sigma v \rangle \iint dv_1\, dv_2\, \sigma_{1,2}(v) v f_l(v_1)$$

where $v = |v_1 - v_2|$ and the integrals are taken over the three-dimensional velocity space. In order to put eqn $\langle \sigma v \rangle \iint dv_1\, dv_2\, \sigma_{1,2}(v) v f_l(v_1)$ in a form suitable for integration, we express the velocities v_1 and v_2 by means of the relative velocity and of the center-of-mass velocity $v_c = (m_1 v_1 + m_2 v_2)/(m_1 + m_2)$:

$$v_1 = v_c + v m_2 / (m_1 + m_2);$$

$$v_2 = v_c - v m_1 / (m_1 + m_2).$$

Equation $\langle \sigma v \rangle \iint dv_1\, dv_2\, \sigma_{1,2}(v) v f_l(v_1)$ then becomes

$$\langle \sigma v \rangle = \frac{(m_1\, m_2)^{3/2}}{(2\pi k_B T\,)^3} \times \iint dv_1\, dv_2\, \exp\left(-\frac{(m_1 + m_2) v_c^2}{2k_B T} - \frac{m_r v^2}{2k_B T}\right) \sigma(v) v,$$

Where m_r is the reduced mass defined by equation $m_r = \frac{m_1 + m_2}{m_1 + m_2}$, and the subscripts '1,2' have been omitted. It can be shown that the integral over $dv_1\, dv_2$ can be replaced by an integral over $dv_c\, dv$, so that we can write

$$\langle \sigma v \rangle \left[\left(\frac{m_1 + m_2}{2\pi k_B T}\right)^{3/2} \int dv_c \exp\left(-\frac{(m_1 + m_2)}{2k_B T} v_c^2\right)\right] \times \left(\frac{m_r}{2\pi k_B T}\right)^{3/2} \int dv_c \exp\left(-\frac{m_r}{2k_B T} v^2\right) \sigma(v) v.$$

The term in square brackets is unity, being the integral of a normalized Maxwellian, and we are left with the integral over the relative velocity. By writing the volume element in velocity space as $dv = 4\pi v^2$, and using the definition of center-of-mass energy $\in$, we finally get

$$\langle \sigma v \rangle = \frac{4\pi}{(2\pi m_r)^{1/2} (k_B T)^{3/2}} \int_0^\infty \sigma(\in) \in exp(-\in / k_B T) d\in.$$

Gamow form for non-resonant Reactions

Useful and enlightening analytical expressions of the reactivity can be obtained by using the simple parameterization $\sigma(\in) = \frac{S(\in)}{\in} \exp(-\sqrt{\in G/\in})$ of the cross-section. In this case the integrand of eqation $\langle \sigma v \rangle = \frac{4\pi}{(2\pi m_r)^{1/2} (k_B T)^{3/2}} \int_0^\infty \sigma(\in) \in exp(-\in / k_B T) d\in.$ becomes

$$y(\in) = S(\in) exp\left[-\left(\frac{\in_G}{\in}\right)^{1/2} - \frac{\in}{k_B T}\right] = S(\in) g(\in, k_B T).$$

An interesting result is obtained for temperatures $T << \in_G$ and stems from the fact that the function $g(\in, k_B T)$ is the product of a decreasing exponential coming from the Maxwellian times an increasing one originating from the barrier penetrability, as shown in below Figure. It has a maximum at the Gamow peak energy

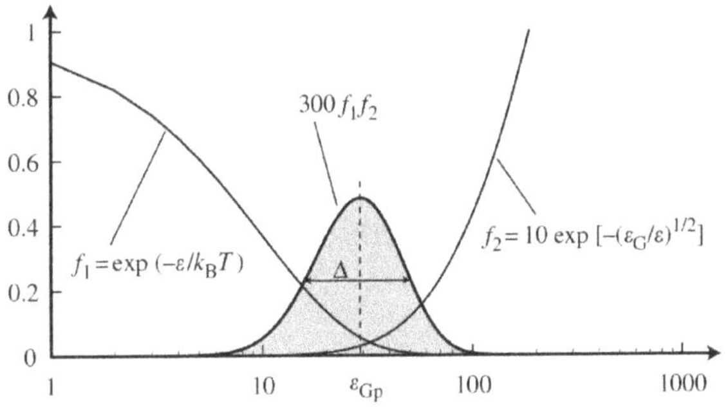

Center-of-mass energy (keV)

$$\in_{Gp} = \left(\frac{\in_G}{4k_B T}\right)^{1/3} k_B T = \xi k_B T,$$

Where, for equation $\in_G = (\pi\alpha_f Z_1 Z_2)^2 2m_r C^2 = 9.86.1 Z_1^2 Z_2^2 A_r$ keV

$$\xi = 6.2696 (Z_1 Z_2)^{2/3} A_r^{2/3} T^{-1/3},$$

with the temperature in kiloelectron volt. To perform the integration we use the saddle-point method, that is, we first expand $y(\in)$ in Taylor series around $\in = \in_{Gp}$, thus writing

$$y(\in) \cong S(\in) \exp\left[-3\xi + \left(\frac{\in - \in_{Gp}}{\Delta/2}\right)^2\right],$$

With

$$\Delta = \frac{4}{\sqrt{3}} \xi^{1/2} k_B T .$$

Equation $y(\in) \cong S(\in) \exp\left[-3\xi + \left(\frac{\in - \in_{Gp}}{\Delta/2}\right)^2\right]$, shows that most of the contribution to the reactivity comes from a relatively narrow energy region with width Δ centered around $\in = \in_{Gp}$, in the high energy portion of the velocity distribution function.

Using equation $y(\in) = S(\in)\, exp\left[-\left(\frac{\in_G}{\in}\right)^{1/2} - \frac{\in}{k_B T}\right] = S(\in) g(\in, k_B T).- \Delta = \frac{4}{\sqrt{3}} \xi^{1/2} k_B T$. and with the further assumption of nonexponential behaviour of $S(\in)$ we can integrate equation $\langle \sigma\, v \rangle \left[\left(\frac{m_1 + m_2}{2\pi k_B T}\right)^{3/2} \int dv_c \exp\left(-\frac{(m_1 + m_2)}{2k_B T} v_c^2\right)\right] \times \left(\frac{m_r}{2\pi k_B T}\right)^{3/2} \int dv_c \exp\left(-\frac{m_r}{2k_B T} v^2\right) \sigma\,(v) v.$ to get the reaction rate in the so-called Gamow form

$$\langle \sigma\, v \rangle = \frac{8}{\pi\sqrt{3}} \frac{\hbar}{m_r Z_1 Z_2 e^2} \bar{S} \xi^2\, exp(-3\xi\,).$$

Here, we have used $\int_0^\infty \exp(-x^2)\, dx = \sqrt{\pi}/2$, and indicated with $\bar{S}$ an appropriately averaged value of S. In the cases in which S depends weakly on $\in$, one can simply set $\bar{S} = S(0)$. In the following, when distinguishing between $\bar{S}$ and $S(0)$ is not essential, we shall simply use the symbol S. Inserting the values of the numerical constants above equation becomes

$$\langle \sigma\, v \rangle = \frac{6.4 \times 10^{-18}}{A_r Z_1 Z_2} S \xi^2\, exp(-3\xi\,)\, cm^3 / s,$$

Where S is in units of kiloelectron volt barn and ξ is given by equation $\xi = 6.2696 (Z_1 Z_2)^{2/3} A_r^{2/3} T^{-1/3}$, We remark that the Gamow form is appropriate for reactions which do not exhibit resonances in the relevant energy range. In particular, it is a good approximation for the DD reactivity, while it is not adequate for the DT and D ^3He reactions.

Above Equation can be used to appreciate the low-temperature behaviour of the reactivity. By differentiation we get

$$\frac{d\langle \sigma\, v \rangle}{\langle \sigma\, v \rangle} = -\frac{2}{3} + \xi \frac{dT}{T},$$

which leads to

$$\langle \sigma\, v \rangle \propto T^{\xi}$$

When $\xi >> 1$. A strong temperature dependence is then found when $T << 6.27 Z_1^2 Z_2^2 A_r$, making apparent the existence of temperature thresholds for fusion burn, which are increasing functions of the mass of the participating nuclei.

Reactivity of Resonant Reactions

When a reaction exhibits a resonance in the energy interval of interest, the astrophysical S factor appearing in the parametrization is a strongly varying factor of energy. As a consequence, the reactivity cannot be expressed in the Gamow form $\langle \sigma v \rangle = \frac{8}{\pi\sqrt{3}} \frac{\hbar}{m_r Z_1 Z_2 e^2} \bar{S} \xi^2 \, exp(-3\xi)$. For a reaction with a single resonance at energy $\in_r$ we can instead use the Breit–Wigner form of the cross section

$$\sigma \propto \lambda\!\!\!^{-}{}^2 \frac{\Gamma_a \Gamma_b}{(\in - \in_r)^2 + (\Gamma/2)^2},$$

Where Γ is the width of the resonance and Γ_a and Γ_b are the so-called partial widths for the input and the output reaction channels. When Γ is sufficiently small, the cross section takes large values in a narrow energy range centered around $\in = \in_r$. In this interval the channel widths can be taken as constants. The relevant Maxwellian reactivity can then be simply evaluated by assuming that only nuclei with energy falling in the resonance peak contribute to it. We thus have

$$\langle \sigma v \rangle \simeq \sigma(\in_r) f(\in_r) v_r \frac{\Gamma}{2} \propto T^{-3/2} \exp\left(-\frac{\in_r}{T}\right), \tag{1}$$

where $f(v)$ is the relevant Maxwellian velocity distribution function and $v_r = (2\in_r / m_r)^{1/2}$.

Reactivities for Controlled Fusion Fuels

Curves of the reactivity as a function of the temperature, obtained by numerical integration of eqn $\left(\langle \sigma v \rangle \left[\left(\frac{m_1 + m_2}{2\pi k_B T} \right)^{3/2} \int dv_c \exp\left(-\frac{(m_1 + m_2)}{2k_B T} v_c^2 \right) \right] \times \left(\frac{m_r}{2\pi k_B T} \right)^{3/2} \int dv_c \exp\left(-\frac{m_r}{2k_B T} v^2 \right) \sigma(v) v \right)$ with the best available cross-sections, are shown in below figure for the reactions of interest to controlled fusion. We see that the DT reaction has the largest reactivity in the whole temperature interval below 400 keV. The DT reactivity has a broad maximum at about 64 keV; it is 100 times larger than that of any other reaction at 10–20 keV and 10 times larger at $50\,\text{keV}$. The second most probable reaction is DD at temperatures $T < 25\,\text{keV}$, while it is D 3He for $25 < 250\ keV$. The reactivity of p 11B equals that of D 3He at temperature about 250 keV and that of DT at about 400 keV. At such very high temperatures other reactions (such as T ^{3}He, p ^{9}Be, D ^{6}Li) have reactivity comparable to that of p ^{11}B, but they are less interesting for controlled fusion because the fuels involved either contain rare isotopes or generate radioactivity.

In the range of temperatures 1–100 keV the reactivity of the DT, DD, and $D\,^3$He reactions are accurately fitted by the functional form

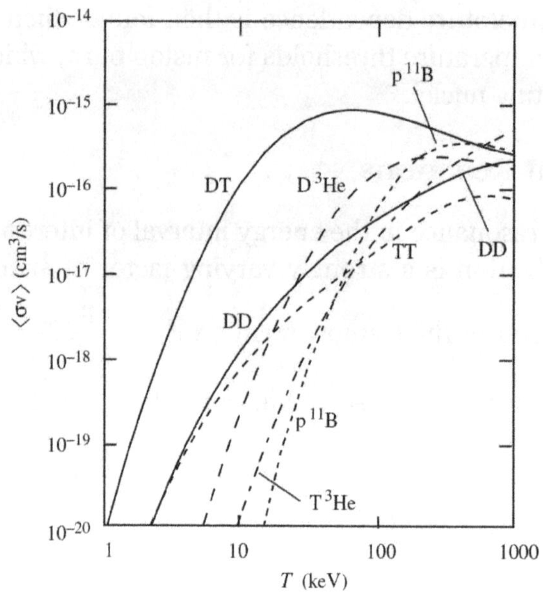

Figure: Maxwell-averaged reaction reactivity versus temperature for reactions of interest to controlled fusion.

$$\langle \sigma v \rangle = C_1 \zeta^{-5/6} \xi^2 \exp\left(-3\zeta^{1/3}\xi\right)$$

Where

$$\zeta = 1 - \frac{C_2 T + C_4 T^2 + C_6 T^3}{1 + C_3 T + C_5 T^2 + C_7 T^3}$$

and ξ is given by equation, which can be written as

$$\xi = C_0 / T^{1/3}.$$

It is seen that the Gamow form $\langle \sigma v \rangle = \dfrac{6.4\times 10^{-18}}{A_r Z_1 Z_2} S\xi^2\, exp(-3\xi)\ cm^3/s,$ is recovered as the temperature $T \to 0$, and then $\zeta \to 1$. At high temperatures the assumed functional form allows for including reactions occurring in the wing of a resonance. The values of the constant C_0 and of the fit coefficients $C_1 - C_7$ appearing in above equation are listed in table. The table also gives estimated errors of the fit.

The reactivity of the $p^{11}B$ reaction, instead, is well fitted by the expression

$$\langle \sigma v \rangle_{pB} = C_1 \zeta^{-5/6} \xi^2 \exp\left(-3\zeta^{1/3}\xi\right) + 5.41\times 10^{-15} T^{-3/2} \times \exp\left(-148/T\right) cm^3/s, \quad (2)$$

with ζ still given by equation $\zeta = 1 - \dfrac{C_2 T + C_4 T^2 + C_6 T^3}{1 + C_3 T + C_5 T^2 + C_7 T^3}$, and the values of the coefficients $C_0 - C_7$ listed in below table. The second term on the right-hand side of above equation accounts for the previously mentioned narrow resonance at 148 keV, and has the functional form (1). In eqn (2) and in the following below equation the temperature T is in units of kiloelectron volt.

Table: Parameters for the reactivity fit, above equation here energies and temperatures are in keV, and the reactivity in cm^3/s. The compact notation $A(b, c)D$ used here stands, as usual, for $A+B \rightarrow C+D$.

Reaction		T(d, n)α	D(d, p)T	D(d, n)^{3}He	^{3}He(d, p)α	^{11}B(p, α)2α
Fit (eqn number)		1.62	1.62	1.62	1.62	1.65
C_0	keV$^{1/3}$	6.6610	6.2696	6.2696	10.572	17.708
$C_1 \times 10^{16}$	cm^3/s	643.41	3.7212	3.5741	151.16	6382
$C_2 \times 10^3$	keV^{-1}	15.136	3.4127	5.8577	6.4192	−59.357
$C_3 \times 10^3$	keV^{-1}	75.189	1.9917	7.6822	−2.0290	201.65
$C_4 \times 10^3$	keV^{-2}	4.6064	0	0	−0.019108	1.0404
$C_5 \times 10^3$	keV^{-2}	13.500	0.010506	−0.002964	0.13578	2.7621
$C_6 \times 10^3$	keV^{-3}	−0.10675	0	0	0	−0.0091653
$C_7 \times 10^3$	keV^{-3}	0.01366	0	0	0	0.00098305
T range	keV	0.2–100	0.2–100	0.2–100	0.5–190	50–500
Error		<0.25%	<0.35%	<0.3%	<2.5%	<1.5%

Simpler formulas are useful for rapid evaluations. For the DT reaction, which is by far the most important one for present fusion research, the expression

$$\langle \sigma v \rangle_{DT} = 9.10 \times 10^{-16}\, exp\left(-0.572 \left|\ln \frac{T}{64.2}\right|^{2.13}\right) cm^3/s\ ,$$

is 10% accurate in the range $3-100$ keV, and 20% accurate in the range 0.3–3 keV. Power law expressions can be useful in analytic studies. In particular, in the temperature range 8–25 keV the DT reactivity is approximated to within 15% by

$$\langle \sigma v \rangle_{DT} = 1.1 \times 10^{-18}\, T^2\, cm^3/s.$$

For the two main branches of the DD reaction good approximations are provided by slightly modified Gamow expressions:

$$\langle \sigma v \rangle_{DDp} = 2 \times 10^{-14} \frac{1 + 0.00577T^{0.949}}{T^{2/3}}\, exp\left(-\frac{19.31}{T^{1/3}}\right) cm^3/s$$

and

$$\langle \sigma v \rangle_{DDn} = 2.72 \times 10^{-14} \frac{1 + 0.00539T^{0.917}}{T^{2/3}}\, \exp\left(-\frac{19.80}{T^{1/3}}\right) cm^3/s.$$

Here the subscripts DDp and DDn indicate the reaction branches $D+D \rightarrow T\ (1.01\ MeV) + p\,(3.03\ MeV)$ and $D+D \rightarrow {}^3He\,(0.82\ MeV) + n\ (2.45\ MeV)$, releasing a proton and a neutron, respectively. Above Equations are about 10% accurate in the temperature range 3–100 keV.

For the D ³He reactions one can use the expression

$$\langle \sigma v \rangle_{D3He} = 4.98 \times 10^{-16}\, exp\left(-0.152\left|\ln\frac{T}{802.6}\right|^{2.65}\right)\, cm^3/s,$$

which is 10% accurate for temperatures in the range 0.5–100 keV

It is interesting to compare the above reactivities to that of the pp reaction.

$$\langle \sigma v \rangle_{pp} = 1.56\times10^{-37}T^{-2/3}\, exp\left(-\frac{14.94}{T^{1/3}}\right)$$
$$\times\left(1 + 0.044T + 2.03 \times 10^{-4}T^2 + 5\times10^{-7}T^3\right)\, cm^3/s.$$

We immediately find that the pp reactivity is 24–25 orders of magnitude smaller than that of DT at temperatures of 1–10 keV. It may be surprising to observe that the specific power of fusion reactions at the center of the Sun takes the very small value of 0.018 W/kg, that is, about 1/50 the metabolic heat of the human body.

Nuclear Fission

The splitting of a heavy nucleus into two smaller fragments of approximately equal mass is known as nuclear fission. There is some mass defect observed during fission, the total mass of the products of fission is less than total mass of the neutron and the Uranium-235 atom. The loss of mass appears in the form of energy according to Einstein's mass energy relation. $E = mc^2$

The nuclear fission of Uraniom-235 is as follow:

$${}_{92}U^{235} + 0n^1 \rightarrow 3{}_0n^1 + {}_{36}Kr^{92} + {}_{56}Ba^{141} + \text{ENERGY}$$

Nuclear Fission

A general nuclear fission reaction involves the fission of heavy nucleus into small nucleus and active subatomic particles which can further attack on another heavy nucleus to continue the chain reaction.

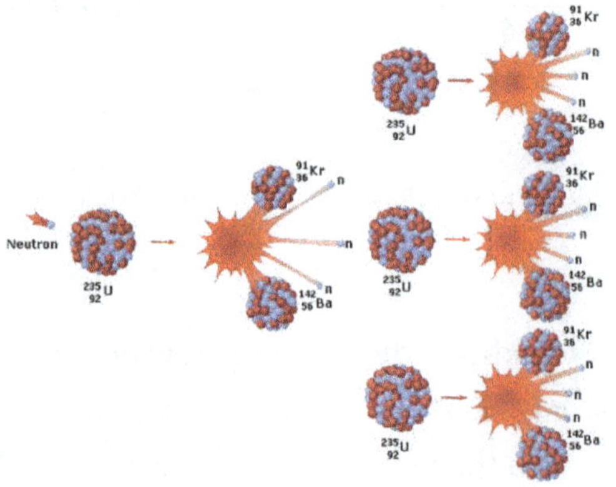

The well-known example of nuclear fission is fission of radioactive isotope Uranium -235 due to the bombardment of thermal neutrons to form barium-114 and krypton-92 with three active neutrons.

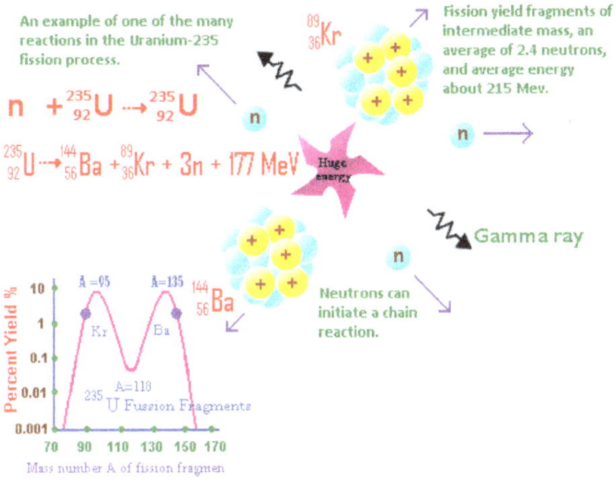

Nuclear Fission Equation

In general in any nuclear fission reaction, a heavy nucleus disintegrated in to one or more small nuclei with some energetic particles. The fission takes place due to the bombardment of some sub-atomic particles like neutrons.

Hence the general nuclear fission equation is

Heavy nucleus + neutrons → Small nuclei (1) + small nuclei (2)+ neutrons + energy

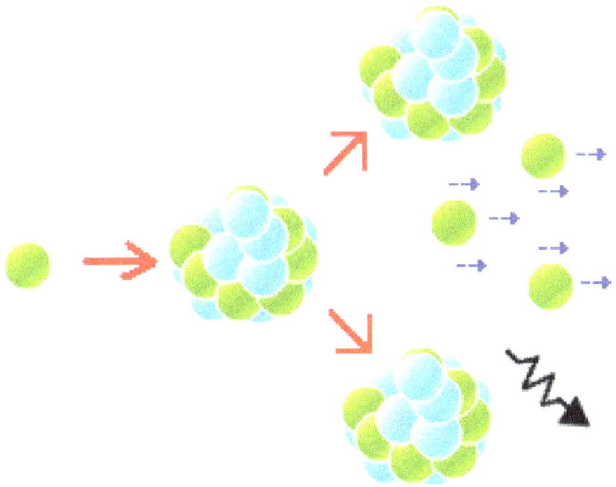

Just like other chemical reactions, nuclear reaction also follow conservation of mass.

For example the nuclear fission of uranium-235 through the bombardment of thermal neutrons.

$$_{92}U^{235} + {_0}n^1 \longrightarrow 3\,{_0}n^1 + {_{36}}Kr^{92} + {_{56}}Ba^{141} + \text{ENERGY}$$

Mass of molecules; 235 +1 = 3x1 + 92 +141 = 236

Examples of Nuclear Fission

Any nuclear fission equation consist of a parent nuclei with neutrons as reactant and form more than one small daughter nuclei with some number of neutrons and large amount of energy.

For example; the nuclear fission of Uranium-235 with one particle of neutron forms Cesium-140, Rubidium-93 and three neutrons and large amount of energy.

$$_{92}U^{235} + _0n^1 \rightarrow _{55}Cs^{140} + _{37}Rb^{93} + 3_0n^1 + \text{Energy}$$

Example of nuclear fission. Some other examples of nuclear fission equations are as follow.

$$92U^{235} + _0n^1 \rightarrow _{56}Ba^{141} + _{36}Kr^{92} + 3_0n^1 + \text{Energy}$$

$$92U^{235} + _0n^1 \rightarrow _{56}Ba^{137} + _{36}Kr97 + 3_0n^1 + \text{Energy}$$

$$92U^{235} + _0n^1 \rightarrow _{92}U^{236} \rightarrow _{54}Xe^{140} + _{38}Sr^{94} + 2_0n^1 + \text{Energy}$$

Nuclear Fission Energy

Some points about nuclear fission energy has given below:

1. Nuclear fission releases a large amount of energy which can be calculated from the binding energies of the nuclei as well as from the mass defect.
2. The nuclear fission of uranium-235 release around 215 MeV of energy.
3. The binding energy of Uranium -235 lies around 7.6 MeV per nucleon while in the case of products of fission lies close to 8.5 MeV per nucleon.
4. While the formation of uranium nucleus from its nucleons can release 235 x 7.6 MeV of energy and the formation of nuclei of products of fission can release about 235 x 8.5 MeV of energy.
5. Hence the energy released during the fission of the nucleus of uranium of the mass number 235 will be equals to 235 (8.5-7.6) MeV = 211.5 MeV.
6. Thus the fission of an atom of releases 211.5 MeV of energy and one mole of uranium-235 released 235 x 6.022 x 1023 MeV = 20.41 x 109 kJ of energy.
7. Hence the energy released during the nuclear fission is million times greater than any chemical reaction.

The same amount of energy can be calculated by using mass defect concept.

$$92U^{235} + _0n^1 \rightarrow _{56}Ba^{141} + _{36}Kr^{92} + 3_0n^1 + \text{Energy}$$

Mass of reactants, 235.044 +1.009 = 236.053 a.m.u.

Mass of products, 91.905 + 140.908 +3.027 = 235.840 a.m.u

Hence the mass defect, *â^ †* M = 236.053-235.840 = 0.231 a.m.u.

Energy released 0.231 a.m.u. x 931.5MeV = 198.41 MeV which is very close to 211.5 MeV.

Energy from Uranium Fission	
Form of energy released	Amount of energy released(MeV)
Kinetic energy of two fission fragments	168
Immediate gamma rays	7
Delayed gamma rays	3-12
Fission neutrons	5
Energy of decay products of fission fragments	---
Gamma rays	7
Beta particles	8
Neutrons	12
Average total energy released	215MeV

Another example if nuclear fission is the fission of Uranium-236 which produced forms uranium-235 by neutron capture, in to Cesium-137 and Rubidium-95 releases around 191.1 MeV of energy.

$$92U^{235} + {}_0n^1 \rightarrow 92U^{236} \rightarrow {}_{55}Cs^{137} + {}_{37}Rb^{95} + 4\,{}_0n^1 + \text{Energy}$$

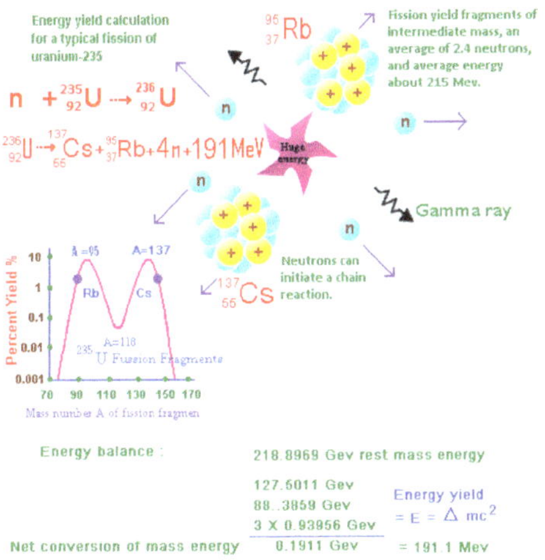

Nuclear Fission Chain Reaction

Nuclear fission is an important nuclear reaction not only because it accompanied by the release of an enormous amount of energy but also because the nuclear reaction resulting from the capture of neutrons which formed as a fission product during the reaction. The neutrons thus released may cause fission of other nuclei of Uranium-235 and set up a chain reaction.

Once initiation required for the reaction, but after that the fission process would be self sustaining with a continuous release of energy. This concept is looking pretty good for the production of large amount of energy, but actually that does not happen.

Hence in order to secure a self-sustaining chain reaction in a fission process, it is necessary to produce a certain number of neutrons, must be at least equals to the number of neutrons involve in fission and non-fission process plus the number of neutrons escapes from the system. Hence we have to increases the surface area of parent nuclei so area by volume ratio decreases, as neutron can escape only through the surface of nuclei so this will decreases the loss of neutrons.

The size of the reactant material which permits the escape of the neutrons to such an extent that at least one neutron is definitely left behind per fission is known as critical size and the corresponding mass is known as critical mass. If the mass of fissionable material is less than critical mass, the fission reaction would not occur and that system is said to be in sub-critical state. If mass is more than critical mass, than only fission reaction would occur and system known as super-critical state.

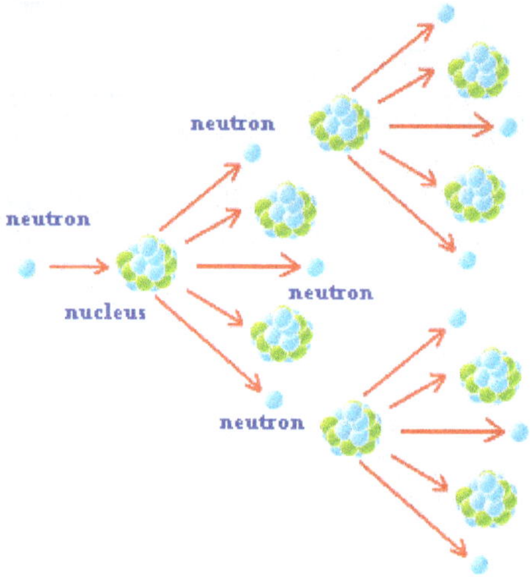

For example: The critical mass for Uranium-235 is 1 kg for an aqueous solution of uranium salt containing 90% of Uranium-235.

Nuclear Fission Reactor

Nuclear fission reactors are device used to produce large amount of energy which can be utilized for some good purpose. These reactors based on the nuclear fission reaction and generally used uranium and polonium isotopes for nuclear fission reaction.

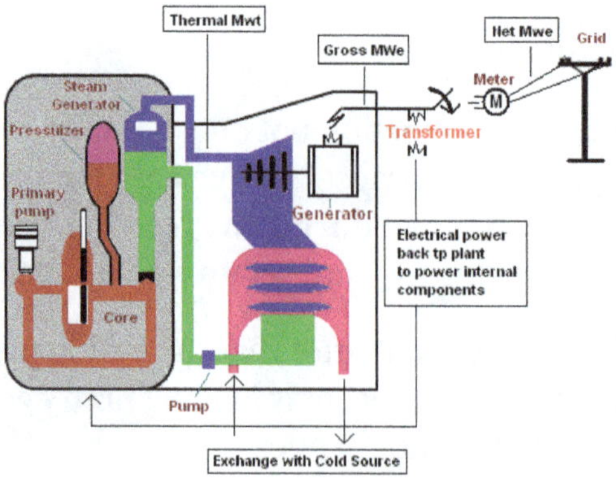

There are several types of nuclear reactor made up of some general components like.

Fuel

- Generally Uranium and polonium used as a basic fuel in nuclear fission reactors.
- Uranium is taken as pellets of uranium oxide (UO_2) which are arranged in tubes to form fuel rods.

Moderator

- There must be some material to decrease the speed of neutron produced as fissionable product.
- These materials used to slow down the speed of neutrons are called as Moderators.
- Generally heavy water and graphite rods serve as a good moderator in reactors.
- Other moderators are helium at 100 atm and 1273 K, beryllium at high temperature, sodium at 773 to 873 K used in breeder reactor or $BeF_2 + ZrF_4$ used in gas-cooled reactors.

Control Rods

- As name implies, these rods used to control the fission reactions.
- They are made up of some neutron-absorbing material like cadmium, hafnium or boron.
- During nuclear fission reaction, they are inserted or withdrawn from the core to control the rate of reaction.

Coolant

- Coolant is a liquid or gas which circulates through the core to transfer the heat from core to coolant.
- Water is a best primary coolant which forms steam after absorbing heat from the core.

Pressure Vessel or Pressure Tubes

These are series of robust steel holding the fuel and conveying the coolant through the moderator.

Steam generator

It is a part of the cooling system where water used as a primary coolant and produce steam for the turbine.

Containment

It is a meter-thick concrete and steel structure around the reactor core which is designed to protect core from outside intrusion as well as to protect those outside from the effects of radiation in case of any malfunction inside.

On the basis of different coolant, moderator and fuel, nuclear fission reactor can be different types.

For example,

Reactor types	Capacity in thousands of megawatts (GWe)	Fuel	Coolant	Moderator
Pressurized water reactor (PWR)	251.6	Enriched UO_2	Water	Water
Boiling water reactor (BWR)	86.4	Enriched UO_2	Water	Water
Pressurized heavy water reactor 'CANDU' (PHWR)	24.3	Natural UO_2	CO_2	Graphite
Gas-cooled reactor (AGR & Magnox)	10.8	Natural U (metal), enriched UO_2	CO_2	Graphite
Light water graphite reactor (RBMK)	12.3	Enriched UO_2	Water	Graphite
Fast neutron reactor (FBR)	1.0	PuO_2 and UO_2	Liquid sodium	none
Other	0.05	Enriched UO_2	Water	Graphite

Advantages of Nuclear Fission

Advantages of Nuclear Fission has Given Below:-

1. In nuclear reactor the nuclear chain reaction is carried out in a controlled manner and liberated heat is converted into electricity.
2. The fuel used in nuclear reactor is relatively expensive and available in trace amounts but fuel used in very little quantity to produce such a large amount of energy.
3. For example; about 28gm of Uranium releases the same amount of energy as the energy released by 100 metric tons of coal.
4. The nuclear fission is not contributes in global warming or other pollution effects which mainly associated with fossil fuel combustion.
5. Because of small quantity required for fuel in nuclear reactors, it can be easily transported.
6. The large heat produced in nuclear reactors can be used to produce cheap electricity which can be further used for other benefits.
7. The waste product formed during the fission process is very less and nuclear reactors are very reliable source of energy.
8. The nuclear reactor can be active for approx. 40 to 60 years.

Advantages and disadvantages of nuclear fission

	Advantages	Disadvantages
1	Produce a large amount of energy	Nuclear power reactor are very compact and capital cost of building is very high
2	Little amount of fuel required	Reactors require high security
3	Less amount of waste product formed	Small leakage from reactor can cause lethal effect
4	No contribution in global warming and pollution	Fuel used in reactors is very expensive and present in trace amount in nature
5	Maintenance and running costs are relatively low	Possibility of uncontrolled nuclear fission can cause nuclear fallout

Uses of Nuclear Fission

- Nuclear fission energy appears promising for space propulsion applications and in the generation of high mission velocities with less reaction mass. This is because of production of large amount of energy, around 106 times more than any chemical reactions and can be used for the current generation of rockets.
- Nuclear fission is generally used in radioisotope thermoelectric generators for space mission.
- Just like conventional thermal power stations which generate electricity by using the thermal energy released from burning fossil fuels, nuclear power plants follow the conversion of the energy released from nuclear fission in nuclear reactor.
- The heat produced during the nuclear fission removes from core by using coolant and used to generate steam which drives a steam turbine. These steam turbines are connected to a generator to produce electricity.

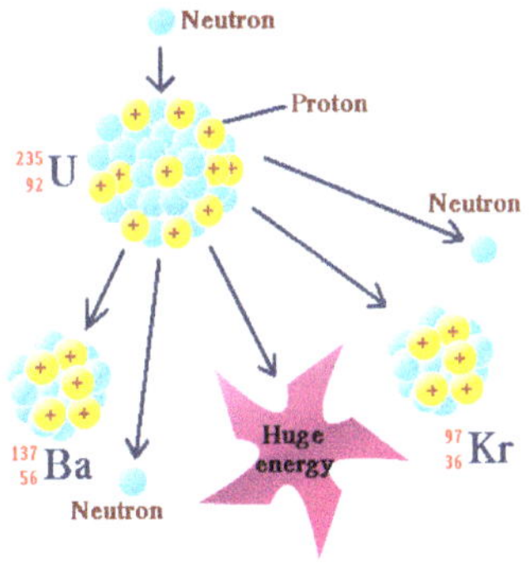

- Nuclear technology is widely used in diagnostics and radiation treatment like radiation therapy for cancer.
- Nuclear fission is used to produce some less common radioisotopes like cesium-137 (Cs-137) by using uranium-235 which is used in photographic sources.
- Nuclear fission energy is also use as a power source for propelling submarines and some type of surface vessels.
- Another application of nuclear fission reactors is their high neutron fluxes which can be used for studying the structure and properties of materials.

Permissions

All chapters in this book are published with permission under the Creative Commons Attribution Share Alike License or equivalent. Every chapter published in this book has been scrutinized by our experts. Their significance has been extensively debated. The topics covered herein carry significant information for a comprehensive understanding. They may even be implemented as practical applications or may be referred to as a beginning point for further studies.

We would like to thank the editorial team for lending their expertise to make the book truly unique. They have played a crucial role in the development of this book. Without their invaluable contributions this book wouldn't have been possible. They have made vital efforts to compile up to date information on the varied aspects of this subject to make this book a valuable addition to the collection of many professionals and students.

This book was conceptualized with the vision of imparting up-to-date and integrated information in this field. To ensure the same, a matchless editorial board was set up. Every individual on the board went through rigorous rounds of assessment to prove their worth. After which they invested a large part of their time researching and compiling the most relevant data for our readers.

The editorial board has been involved in producing this book since its inception. They have spent rigorous hours researching and exploring the diverse topics which have resulted in the successful publishing of this book. They have passed on their knowledge of decades through this book. To expedite this challenging task, the publisher supported the team at every step. A small team of assistant editors was also appointed to further simplify the editing procedure and attain best results for the readers.

Apart from the editorial board, the designing team has also invested a significant amount of their time in understanding the subject and creating the most relevant covers. They scrutinized every image to scout for the most suitable representation of the subject and create an appropriate cover for the book.

The publishing team has been an ardent support to the editorial, designing and production team. Their endless efforts to recruit the best for this project, has resulted in the accomplishment of this book. They are a veteran in the field of academics and their pool of knowledge is as vast as their experience in printing. Their expertise and guidance has proved useful at every step. Their uncompromising quality standards have made this book an exceptional effort. Their encouragement from time to time has been an inspiration for everyone.

The publisher and the editorial board hope that this book will prove to be a valuable piece of knowledge for students, practitioners and scholars across the globe.

Index

www.ingramcontent.com/pod-product-compliance
Lightning Source LLC
Chambersburg PA
CBHW082046190326
41458CB00010B/3468
9781641721103